organic & natural

the website guide

ORGANIC & NATURAL

THE WEBSITE GUIDE

compiled by

Louis Douglas

WEBSITE GUIDES

Version 1.1

First published in Great Britain in 2002 by
Website Guides
PO Box 132
Ipswich
IP6 9PH

editorial@websiteguides.co.uk

www.websiteguides.co.uk

ISBN 1 903755 02 6

Printed and bound in Great Britain
by the Lavenham Press

Contents

INTRODUCTION

The web is a wonderful tool which gives people with computers, wherever they live and however busy they are, almost instant access to a huge amount of useful information, goods and services, much of it free. The only drawback can be knowing how to find the sites you really want.

Website Guides are designed to overcome this problem with practical, informative and useful special interest guides to the internet.

Each guide covers a particular subject area. Comprehensive details of relevant sites have been sifted in advance, informatively indexed and clearly presented . Each entry describes what you will find on the site and gives full information on specialities, on-line ordering, email, contact, address, telephone and fax. The subject indexes are wide-ranging and easy to use. They provide a fast-track way to locate those sites which you know will be of interest before you sit down at your computer.

Search engines, on the other hand, can often be time-consuming and potentially unrewarding. You may be lucky to stumble on sites of interest or you may spend tedious hours logging on to promising sounding names which turn out to be of no interest. If you have loads of time to spare, this can be fun and lead you down unexpected paths. But if time is limited search engines can prove to be both cumbersome and slow. Website guides. on the other hand, give you instant information at your fingertips.

In contrast to many other internet directories in book form, Website Guides have two key features which make them particularly practical and user-friendly. Firstly, in the business and service listings, full details of name, contact, email, address, telephone and fax numbers are given wherever possible. These make the guides proper working directories and important works of reference. Secondly the comprehensive subject and other indexes allow businesses and organisations to be listed and cross referenced under a number of different subject areas rather than under just one heading.

Website Guides are intended for busy, computer literate people who want to make the best use of their time. They have specific interests, such as gardening or organics, which they want to follow up. They are people who use the web as a tool to obtain useful information and services or to purchase goods on-line.

Websites are continually changing and the information given is correct at the time of writing. Inevitably there will be some changes. Sites come and go, are updated or even change their names. No guide can ever be wholly complete but we believe that the extensive coverage in Website Guides will provide a solid basis for acquiring information and providing jumping off points for further finds.

This particular Guide covers entries under the broad heading Organic and Natural. The term organic can only be used where the grower or manufacturer holds a recognised Certificate of Registration. This would apply to many, but not all of the businesses listed in this book. In addition to strictly organic entries there are also many for businesses and organisations which, whilst not strictly organic, do set out to be environmentally friendly in their approach and subscribe to the principles of a natural and sustainable lifestyle.

Website Guides

Website Guides will be regularly revised and brought up to date. Suggestions for new sites to be included in revised editions, as well as any corrections or updates to existing entries, are always very welcome. Please email or write to

editorial@websiteguides.co.uk
Website Guides - PO Box 132 - Ipswich - IP6 9PH

For information on titles available and forthcoming please visit our website www.websiteguides.co.uk or contact us direct.

Most of the information contained in this guide has been sourced from the websites themselves. Whilst every care has been taken to ensure accuracy, Website Guides cannot take responsibility for any errors, misrepresentations or omissions which may have occurred.

Inclusion of a website in Website Guides does not carry with it any endorsement of quality or service. In some cases, especially where health is concerned, there is an organisation or association covering the particular subject area and their websites can be consulted for further verification.

Websites are included and described on the basis that they are relevant to the subject area of the guide and are or will shortly be operational at the time of writing.

ACKNOWLEDGEMENT

Website Guides would like to thank Huggababy Natural Baby Products and Tony Morrison for the use of photographs on the cover of this book.

The compiler would also like to thank Alison Hart and Louis James for their assistance.

Subject Index

Animals & Pets

animal care
- British Association of Homoeopath Veterinary Surgeons
- Denes Natural Pet Care
- Green Ark Animal Nutrition
- Herbaticus
- Hilton Herbs

animal feed
- Hi Peak Feeds
- The Organic Feed Company
- Vitrition

dog food
- Hi Peak Feeds

food supplements for animals
- Herbaticus

natural remedies for pets
- By Natural Selection

pet food
- Denes Natural Pet Care
- Green Ark Animal Nutrition
- Organic Connections Int.
- Pascoes
- Pero Foods
- Wafcol

pet insurance
- Animal Friends Insurance

petcare
- British Association of Homoeopath Veterinary Surgeons
- By Natural Selection
- Denes Natural Pet Care

veterinary
- British Association of Homoeopath Veterinary Surgeons

Business & Finance

banking
- The Co-Operative Bank
- Triodos Bank

building society
- Ecology Building Society
- Shared Interest Society

business to business
- Organics-on-Line

buying group
- The Health Store

distributors
- Brewhurst Health Food Supplies
- Essential Trading
- Gordon Jopling Food Ingredients
- Grafton Int.
- Savant Distribution
- H Uren & Sons

economics
- New Economics Foundation

ethical choices
- Ethical Junction

ethical investment
- Ethical Financial
- The Ethical Investors Group
- Ethical Money Ltd
- The Ethical Partnership
- Holden Meehan IFA
- UK Social Investment Forum

financial matters
- The Co-Operative Bank
- Ethical Financial
- The Ethical Investors Group
- Ethical Junction
- Ethical Money Ltd
- The Ethical Partnership
- Holden Meehan IFA
- ICOF
- Naturesave Policies
- New Economics Foundation
- Shared Interest Society
- Triodos Bank
- UK Social Investment Forum

food industry suppliers
- Alembic Products
- The Billington Food Group
- Blendex Food Ingredients
- Confoco UK
- Cotswold Health Products
- Cotswold Speciality Foods
- Gordon Jopling Food Ingredients
- Holgran
- Guy Lehmann SFI
- Norgrow Int.
- A R Parkin Limited
- Rasanco
- Regency Mowbray Company
- H Uren & Sons

insurance
- Naturesave Policies

internet trading site
- Organics-on-Line

loan finance
- ICOF

printing
- Recycled Paper Supplies

retail information system
- Intecam

show organisers
- Full Moon Communications

stationery
- Recycled Paper Supplies

wholesalers
- Alara Wholefoods
- Big Oz Industries
- Brewhurst Health Food Supplies
- Delfland Nurseries
- Enzafruit Worldwide
- Essential Trading
- Fargro
- Good Food Distributors
- G's Marketing
- The Health Store
- Infinity Foods
- Manor Farm Organic Milk
- Natures Store
- New Farm Organics
- The Organic Herb Trading Co
- The Organic Wholesale Company
- Organictrade
- Pure Organic Food
- Queenswood Natural Foods
- Rush Potatoes
- Suma Wholefoods
- Thames Fruit
- Welsh Hook Meat Centre

conservation
British Trust for Conservation Volunteers
English Nature
Horticultural Correspondence College
Rare Breeds Survival Trust
Schumacher UK
The Wildlife Trusts
The Woodland Trust
WWF Global Environment Network

eco park
The Earth Centre
Eden Project

ecology
The Ecological Design Association
Ecologist
Resurgence Magazine

environment
Centres For Change
Earthwatch
Eden Project
Environ
Environment & Health News
Ethical Consumer
Friends of the Earth
Green Futures
Green Network
Greenpeace
National Association of Environmental Education
Soil Association
Sustain
Women's Environmental Network
Young People's Trust for Environment

genetic engineering
Genetic Engineering Network
The Genetics Forum
Genewatch UK

magazine
Ecologist
Environment & Health News
Green Futures
Resurgence Magazine
Viva

memorial stones
Alternative Memorials

natural death
Alternative Memorials
Greenfield Coffins
Green Woodland Burial Services

sustainable development
Centre for Alternative Technology
Centres For Change
Green Futures
Greenpeace
Schumacher UK
Soil Association
Young People's Trust for the Environment

theme park
The Earth Centre

wildlife
English Nature
The Wildlife Trusts

woodland
The Woodland Trust

woodland burial
Green Woodland Burial Services

alternative energy
British Wind Energy Association
Centre for Alternative Technology
Ecotricity
Electric Car Association
Energy Development Co-operative
Energy Saving Trust
Fibrowatt
Freeplay Energy
Napiers Herbs
National Energy Foundation
Natural Collection
Solar Century
Solar Sense
Solarsaver
Solartwin
SunDog
Super Globe
Solar Design Company
Unit Energy
Wind and Sun

architects
Architype
Bill Dunster Architects
Constructive Individuals
David Clarke Associates
Gale & Snowden
Procter Rihl
Sarah Wrigglesworth Architects

architectural design
Walter Segal Self Build Trust

architectural salvage
Salvo

building advice
Association for Environment Conscious Building

building materials
Architectural Salvage Index
Chauncey's
Construction Resources
Ecomerchant
Environmental Construction Products
Excel Industries
Natural Building Technologies
Salvo
The Swedish Window Company
Ty-Mawr Lime

design
The Ecological Design Association

ecological housing
Hockerton Housing Project
The Eco House

electric cars
Alternative Vehicles Technology
Electric Car Association

energy
Energy Conservation & Solar Centre
Energy Saving Trust
National Energy Foundation
Solar Trade Association

energy saving
The Green Shop

flooring
Chauncey's
DLW Residential Floorings
Fired Earth
Forbo-Nairn
Health Flooring Network
Natural Flooring Direct
The Original Seagrass Company
Solid Floor
Waveney Rush Industry

housing
The Eco House
Ecology Building Society
The Empty Homes Agency
Hockerton Housing Project

insulation
Excel Industries
Natural Building Technologies

lime products
Ty-Mawr Lime

linoleum
DLW Residential Floorings
Forbo-Nairn

natural death
Greenfield Coffins

paint removal
Eco Solutions
Ray Munn

paints
Auro Organic Paint Supplies
Casa Paint Company
Clearwell Caves
Construction Resources
Ecomerchant
Environmental Construction Products
The Green Shop
The Healthy House
Lakeland Paints
Livos Natural Paints
Natural Building Technologies
Nutshell Natural Paints
Ray Munn
Textures
Ty-Mawr Lime

pallets
Pallet Display Systems

recycling
Blackwall
Community Recycling Network
Henry Doubleday Research Association
Magpie Home Delivery
National Recycling Forum
Pallet Display Systems
Recoup
Recycled Paper Supplies
Waste Watch
Water Dynamics

reed and clay boards
Natural Building Technologies

self powered energy products
Freeplay Energy

self-build
Association of Self Builders
Constructive Individuals
Walter Segal Self Build Trust

solar energy
Construction Resources
Energy Development Co-operative
Solar Century
The Solar Design Company
Solar Sense
Solarsaver
Solartwin
SunDog
Wind and Sun

straw bale buildings
Amazon Nails Straw Bale Building
Straw Bale Building Association

water saving
Water Dynamics

wind power
British Wind Energy Association
Energy Development Co-operative
Solarsaver
SunDog)
Wind and Sun

windows
The Swedish Window Company

allergy-free products
The Healthy House

baby bedding
Green Baby

baby buggies
Mountain Buggy

baby carriers
Better Baby Sling
Hippychick
Huggababy Natural Baby Products
Kids in Comfort
Wilkinet Baby Carrier

baby clothes
Born
Eco Babes
Gossypium.co.uk
Green Baby
Greenfibres Organic Textiles
Greensleeves Clothing
Natural Collection
Perfectly Happy People
Planet Vision.co.uk
Spirit of Nature
Su Su Ma Ma World Wear
The Totally Baby Shop.com

baby food
Baby Organix (Organix Brands)
Boots Direct
Cooks Delight
Green and Organic
Hipp Organic
The Organic Baby Food Company
The Organic Shop (Online)
Organico
Sainsbury's
Simply Organic
Simply Organic Food Company
Tesco
The Totally Baby Shop.com
Waitrose

baby products
Earthwise Baby
Lollipop Children's Products
Organics Direct
Sam-I-Am
The Totally Baby Shop.com

baby shoes
Hippychick

baby toiletries
Green Baby

babycare
Beaming Baby.com
Bodywise (UK)
Born
Cheeky Rascals
Country Life
Eco Babes
Ecobaby
ecozone.co.uk
Ethos
The Healthy House
Hippychick
Honeyrose Products
Huggababy Natural Baby Products
Jurlique
La Leche League
Lavera UK
Little Green Earthlets
The Nappy Shop
National Association of Nappy Services
Natural Woman
Nutricia
The Organic Wool Company
Perfectly Happy People
Real Nappy Association
Revital Stores
Tisserand Aromatherapy
Weleda

back trouble
Back in Action

bakery
Infinity Foods
Phoenix Community Stores

bedding
Eco Clothworks
Greenfibres Organic Textiles
Textures

birth choices *(see also water births)*
Association for Improvements in the Maternity Services

birthing pools
Active Birth Centre
Splashdown Water Birth Services

board games
Words of Discovery

breast feeding
Expressions Breastfeeding
La Leche League

chairs
Back in Action

childbirth
Active Birth Centre
The Birth Centre

childcare
Hippychick
Vaccination Awareness Network UK

children's books
Barefoot Books
Letterbox Library
Words of Discovery

children's clothing
Soup Dragon
Su Su Ma Ma World Wear

children's club
Kids Organic Club

cleaning products
Earth Friendly Products
Ecover
Global Eco
Natural Eco Trading

clothing
Cooks Delight
Eco Clothworks
Ganesha
Gossypium.co.uk
Greenfibres Organic Textiles
The Healthy House
Hemp Union
Mother Hemp
Natural Collection
Natural Instincts
Planet Vision.co.uk
Spirit of Nature
Textures
Tucano
Veganstore

cosmetics
Cosmetics To Go
Country Life
Culpeper Herbalists
Healthquest- Organic Blue
Hemp Union
Lavera UK
Potions & Possibilities
Power Health Products

dyes
Colouring Through Nature

fabrics
Greenfibres Organic Textiles
Mother Hemp
Natural Instincts
Textures

fancy dress
Soup Dragon
Su Su Ma Ma World Wear

fleeces
Freerangers

footwear
Conker Shoe Co
Ethical Wares
Freerangers
Green Shoes
Veganline
Veganstore

furnishings
Ganesha

furniture
Ecopine
Harvest Forestry
Mufti
Trannon
Treske

games
Gaia Distribution
Green Board Game Company
Soup Dragon

gifts
Oxfam Fairtrade
Supernature.org
Wooden Wonders

hammocks
The Mexican Hammock Co

hemp products
Hemp Union
Mother Hemp

household products
The Better Food Co
Earth Friendly Products
Eco Babes
Ecover
ecozone.co.uk
The Fresh Food Company
The Green Shop
Greenfibres Organic Textiles
The Healthy House
Myriad Organics
Natural Collection
Natural Eco Trading
Rockland Corporation - TRC
Spirit of Nature
Supernature.org
21st Century Health
Veggi Wash Food Safe

incense
Lotus Emporium

Indian goods
Ganesha

insect repellent
Nature's Dream

introduction agencies
Natural Friends
VMM (Vegetarian Matchmakers)

lifestyle
Network Organic

magazine
Ethical Consumer

maternity
Born
Expressions Breastfeeding
Hipp Organic
Huggababy Natural Baby Products
La Leche League
Mothernature

maternity products
Green Baby

Mexican goods
The Mexican Hammock Co

nappies
Bambino Mio
Beaming Baby.com
Born
Cheeky Rascals
Earthwise Baby
Eco Babes
Eco Clothworks
Ecobaby
Ellie Nappy Company
Ethos
Green Baby
Greenfibres Organic Textiles
Little Green Earthlets
Lollipop Children's Products
The Nappy Lady
The Nappy Shop
National Association of Nappy Services
The Natural Baby Company
Perfectly Happy People
Real Nappy Association
Sam-I-Am
Spirit of Nature
Women's Environmental Network

nappy service
Cotton Bottoms

parenting
Earthwise Baby

perfumes
Cosmetics To Go

pushchairs
Pegasus Pushchairs

rush-weave
Waveney Rush Industry

sanitary products
Bodywise (UK)
Natural Woman

self-powered energy products
Freeplay Energy

soaps
Earthbound
Simply Soaps

tampons
Bodywise (UK)
Natural Woman

textiles
Eco Clothworks
Gossypium.co.uk
The Organic Wool Company

toiletries
Grafton Int.
Neal's Yard Remedies
NHR Organic Oils
Revital Stores
Scullions Organic Supplies
Weleda

toothpaste
Kingfisher Natural Toothpaste

towels
Textures

toys
Green Baby

vegan
Earth Friendly Products
Freerangers
Green Shoes
Plamil Organics
Simply Soaps
Veganline
Veganstore

vegetarian & vegan
Cosmetics To Go
Honesty Cosmetics
Kingfisher Natural Toothpaste
Natural Collection
VMM (Vegetarian Matchmakers)

water birth
Active Birth Centre
Association for Improvements in the Maternity Services
The Birth Centre
Splashdown Water Birth Services

water sculptures
Flow forms

woodcraft
Wooden Wonders

wooden toys
Dawson & Son Wooden Toys
Huggababy Natural Baby Products
Soup Dragon

wool
The Organic Wool Company

Australian cereals
Big Oz Industries

baking ingredients
British Bakels
Holgran
Mrs Moons

balsamic vinegar
Divine Wines

beer
The Beer Shop
The Caledonian Brewing Co
Distinctive Drinks Company
Divine Wines
Rose Blanc Rouge
St Peter's Brewery
Sedlescombe Vineyard
Utobeer
Vinceremos Wines & Spirits
Vintage Roots

biscuits
Tods of Orkney

box schemes
Abel & Cole
Ashfield Organic Farm
The Better Food Co
Bumblebee
Ellis Organics
D W & C M Evans
Fieldfare Organics
The Fresh Food Company
A & D Gielty
Greenwich Organic Foods
Half Moon Healthfoods
Howbarrow Organic Farm
Les Jardiniers du Terroir
Magpie Home Delivery
North East Organic Growers
Organic and Natural Foods
Organic Connections Int.
The Organic Delivery Company
Organic Kosher Foods
The Organic Shop (Online)
Organic Trail
Organics Direct
Organics To Go
Pertwood Organics Co-operative
Pillars of Hercules Farm Shop
Pimhill Farm
Pure Organic Food
Riverford Organic Vegetables
Slipstream Organics
Tamarisk Farm
Urban Organics
Westcountry Organics
Woodlands Farm

bread
Canterbury Wholefoods
Daily Bread Co-operative
Doves Farm Foods
Ellis Organics
Myriad Organics
The Village Bakery

bread mixes
G R Wright & Sons

breakfast foods
Shepherdboy

brewery
The Caledonian Brewing Co

brewing equipment
The Beer Shop

butters
Markus Products

cake mixes
British Bakels

cakes
Mrs Moons

cane sugars
The Billington Food Group

canned goods
Whole Earth Foods

caviar alternative
Finlay's Foods

cereals
Alara Wholefoods
Big Oz Industries
Doves Farm Foods
Jordans
New Farm Organics
Organictrade
Whole Earth Foods

charcuterie
Swaddles Green Farm

cheese
Bumblebee
D W & C M Evans
Irma Fingal-Rock
Markus Products
Sussex High Weald Dairy Products
Teddington Cheese
Westcountry Organics
Wyke Farms

children's food
Miniscoff
Swaddles Green Farm

chocolate
Buxton Foods
The Day Chocolate Company
Green and Black's
Lessiter's
Montezuma's Chocolates
NHR Organic Oils
Organic Days
Organico
Plamil Organics
Rococo Chocolates
Traidcraft plc
Viva

Christmas puddings
Get Real Organic Foods

chutneys
Forest Products
Tracklements

cider
Aspall
Crone's
Sedlescombe Vineyard
Vintage Roots
Westons
Wino Organic Wines

coffees
Brian Wogan
Clipper Teas
Equal Exchange
The Hampstead Tea and Coffee Co
Lunn Links Kitchen Garden
Taylors of Harrogate
The Tea and Coffee Plant

community store
Phoenix Community Stores

confectionery
First Quality Foods
Shepherdboy

cookery courses
Ballymaloe Cookery School
Greencuisine
Little Salkeld Watermill
Mrs Tee's Wild Mushrooms
The Natural Cookery School
Vegetarian Cookery Courses
The Village Bakery
Wing of St Mawes

cordials
Belvoir Fruit Farms
Divine Wines
N & J Mawson
Thorncroft
Vinceremos Wines & Spirits

culinary herbs
Barwinnock Herbs
Jekka's Herb Farm

curries
archiamma
Go Organic

dairy
Daily Bread Co-operative
Fieldfare Organics
Futura Foods UK Limited
Graig Farm Organics
Green and Organic
Horizon Organic Dairy
Meat Matters
Provender Delicatessen
Rachel's Organic Dairy
Riverford Organic Vegetables
Scullions Organic Supplies
Sharpham Partnership
Sussex High Weald Dairy Products
Taste Connection
Wigham Farm
Windmill Hill City Farm Shop
Wyke Farms

dairy-free
Body and Soul Organics
Buxton Foods
Plamil Organics

delicatessen
Fresh & Wild
Provender Delicatessen

desserts
Get Real Organic Foods

dressings
Tracklements

dried fruits
Alara Wholefoods
Country Products
Organictrade
Tropical Wholefoods

dried herbs
Hambleden Organic Herb Trading Co

dried products
Organic Days

eating
Slow Food

eggs
Martin Pitt Freedom Eggs
J Savory Eggs

Windmill Hill City Farm Shop

farm shops
Ashfield Organic Farm
Deverill Trout Farm
D W & C M Evans
A & D Gielty
Napiers Herbs
The Organic Beef Co
The Organic Farm Shop
Organic Roots
Pillars of Hercules Farm Shop
Pimhill Farm
Riverford Organic Vegetables

farmers' markets
Hampshire Farmer's Market
London Farmers' Markets
National Farmers' Market Association

fish
Club Chef Direct
Cornish Fish Direct
Graig Farm Organics
Green and Organic
Hawkshead Organic Trout Farm
Meat Matters
Organic Kosher Foods
Wing of St Mawes

flour
Buxton Foods
Daily Bread Co-operative
Doves Farm Foods
Little Salkeld Watermill
F W P Matthews
Tamarisk Farm
The Village Bakery
G.R. Wright & Sons

food dehydrators
Mayfield Services

food hygiene
Veggi Wash Food Safe

frozen foods
Pure Organics

frozen fruit
Oerleman's Foods UK

frozen vegetables
Oerleman's Foods UK
Provender Delicatessen

frozen vegetarian foods
Goodlife Foods

frozen yoghurt
Cream O'Galloway Dairy Co
Rocombe Farm Fresh Ice Cream

fruit
Abel & Cole
Beano Wholefoods
The Better Food Co
Body and Soul Organics
Canterbury Wholefoods
Daily Bread Co-operative
Ellis Organics
Farm-A-Round
Fieldfare Organics
The Fresh Food Company
Graig Farm Organics
Green and Organic
Marchents
Meat Matters
Myriad Organics
North East Organic Growers
Organic and Natural Foods
Organic Connections Int.
The Organic Delivery Company
Organics Direct
Scullions Organic Supplies
Sundance Market
Thames Fruit
Total Organics

fruit sorbets
Rocombe Farm Fresh Ice Cream

fruit teas
Clipper Teas
London Fruit & Herb Company
Only Natural Products

fruit wines
Broughton Pastures Organic Wines

game
Northfield Farm
Organic Meats & Producers Scotland
Somerset Farm Direct
Welsh Hook Meat Centre
Yorkshire Game

garlic
Garlic Genius

geese
Goodman's Geese
Providence Farm Organic Meats
Somerset Levels Organic Foods

gluten-free
Body and Soul Organics
Bumblebee
Country Products
Plamil Organics
The Shieling
Unicorn Grocery

goat dairy products
Naturemade

gourmet home delivery service
Leapingsalmon.com

grains
Daily Bread Co-operative

groceries
Abel & Cole
The Better Food Co
Food Revolution
Greenwich Organic Foods
Harvest Forestry
Home Farm Deliveries
On The Eighth Day Co-operative
The Organic Delivery Company
The Organic Farm Shop
The Organic Shop (Online)
Provender Delicatessen
Sainsbury's
Scullions Organic Supplies
Sundance Market
Taste Connection
Tesco
Total Organics
Unicorn Grocery
Veganstore
Waitrose

halal
Organic Kosher Foods

hampers
Organic Oxygen

hampers for children
Natures Cocoons

hemp foods
Hemp Food Industries Association
Hemp Union

herbal teas
London Fruit & Herb Company
Naturex
Only Natural Products

herbs
Bart Spices
Hambleden Organic Herb Trading Co
Lunn Links Kitchen Garden
North East Organic Growers

home delivery
Allergyfree Direct
Beano Wholefoods
Body and Soul Organics
Farm-A-Round
Food Revolution
Green and Organic
Home Farm Deliveries
Magpie Home Delivery
Organic and Natural Foods
Organic Connections Int.
The Organic Shop (Online)
Organics To Go
Sainsbury's
Simply Organic Food Company
Sundance Market
Sundrum Organics
Tesco
Waitrose
Westcountry Organics

honey
Cotswold Speciality Foods
New Zealand Natural Food Company
Rowse Honey
Traidcraft plc
Tropical Forest Products

ice cream
Cream O'Galloway Dairy Co
First Foods
Mother Hemp
Provender Delicatessen
Rocombe Farm Fresh Ice Cream
September Organic Dairy

Indian food
Organic India

Italian foods
Danmar Int.
Take It From Here

jams
Forest Products
Organic Days
Thursday Cottage
The Village Bakery

Japanese foods
Clearspring
On The Eighth Day Co-operative

juice bar
Fresh & Wild
Pause
Sauce Organic Diner

juicers
Organics Direct

juices
Aspall
Crone's
James White
N & J Mawson
The Organic Wine Company
Organico
Provender Delicatessen
Rose Blanc Rouge
Sedlescombe Vineyard
Vintage Roots
Wino Organic Wines

ketchup
Seeds of Change

kosher foods
Finlay's Foods
Organic Kosher Foods

local produce nationwide
BigBarn

macrobiotic foods
Clearspring
On The Eighth Day Co-operative

magazine
Full Moon Communications
Organic Living

maple syrup
Rowse Honey

markets
Women's Institute Country Markets

meat
Abel & Cole
The Better Food Co
Body and Soul Organics
Carmichael Estate Farm Meats
Chops Away
The Cotswold Gourmet
Dittisham Fruit Farm
Eastbrook Farm Organic Meats
Ellis Organics
M Feller Organic Butchers
Fieldfare Organics
The Fresh Food Company
Graig Farm Organics
Green and Organic
Greenstuff
Growing Concern Organic Farm
Heal Farm Meats
Higher Hacknell Farm
Highland Organic Foods
Howbarrow Organic Farm
Lloyd Maunder
Marchents
Meat Matters
Munson's Poultry
Myriad Organics
New Farm Organics
Northfield Farm
Old Macdonalds
Organic and Natural Foods
The Organic Beef Co
The Organic Farm Shop
Organic Kosher Foods
Organic Meats & Producers Scotland
The Organic Shop (Online)
Providence Farm Organic Meats
Pure Organic Food
Richard Guy's Real Meat Company
Scullions Organic Supplies
Simply Organic Food Company
Somerset Farm Direct
Somerset Levels Organic Foods
Sundance Market
Swaddles Green Farm
Tamarisk Farm
Taste Connection
Thorogoods (Organic Meat)
Total Organics
Welsh Hook Meat Centre
Westcountry Organics
Wigham Farm
Windmill Hill City Farm Shop
Woodlands Farm

milk
Manor Farm Organic Milk
Organic Matters

muesli
Alara Wholefoods
Country Products

mushrooms
Mrs Tee's Wild Mushrooms

mustards
Forest Products
Tracklements

non-dairy
Clearspring
First Foods

nuts
Alara Wholefoods
Organictrade

oatcakes
Nairn's Oatcakes
Tods of Orkney

olive oils
Bumblebee
Danmar Int.
Divine Wines
Irma Fingal-Rock
Organico
Take It From Here
Taste Connection
Vinceremos Wines & Spirits

pasta
Buxton Foods
Clearspring
First Quality Foods
Seeds of Change
Take It From Here

pasta sauces
Simply Organic

pies
Get Real Organic Foods

pizza
Pizza Organic

potato chips
Kettle Organics & Kettle Chips

poultry
Ellis Organics
Heal Farm Meats
Howbarrow Organic Farm
Kelly Turkeys
Lloyd Maunder
Meat Matters
Munson's Poultry
Northfield Farm
Organic Kosher Foods
Organic Meats & Producers Scotland
Providence Farm Organic Meats
Pure Organic Food
Richard Guy's Real Meat Company
Somerset Farm Direct
Somerset Levels Organic Foods
Swaddles Green Farm
Thorogoods (Organic Meat)
Welsh Hook Meat Centre
Wigham Farm
Windmill Hill City Farm Shop

prepared vegetables
Farm-A-Round

preserves
Forest Products

pulses
New Farm Organics
Organictrade

raw food
Fresh Network

ready meals
Café@Yum
Go Organic
Growing Concern Organic Farm
Kidz Organic Kitchen
Leapingsalmon.com
Lloyd Maunder
Organic India
Organico
Oscar Mayer
Pure Organics
Swaddles Green Farm

ready meals for children
Miniscoff

rooibos tea
Wistbray

salt
Anglesey Sea Salt Co

sauces
Alembic Products
Go Organic
Meridian Foods
Seeds of Change
Tracklements

seafood
Marchents

sheep cheeses
Buxton Foods

sheep dairy products
Naturemade

sheep's milk
Sussex High Weald Dairy Products

shellfish
Loch Fyne Oysters

shitake mushrooms
Jac by the Stowl

shortbread
Tods of Orkney

smoked fish
Coln Valley Smokery
Lechlade Trout Farm

smoked salmon
Club Chef Direct
Coln Valley Smokery
Kinvara Smoked Salmon
Loch Fyne Oysters

smoothies
Innocent
Pete & Johnny's PLC
Soma Organic Smoothies

snacks
Jordans
Kettle Organics & Kettle Chips
Lyme Regis Fine Foods

soft drinks
Thorncroft
Westons
Whole Earth Foods

soft fruit
Howbarrow Organic Farm
OrganicSeeds.co.uk

soups
Go Organic
Seeds of Change
Simply Organic

special diets
Allergyfree Direct
GoodnessDirect
Potter's Herbal Medicines

spices
archiamma
Bart Spices
Cotswold Health Products
Country Products
Lunn Links Kitchen Garden
A R Parkin Limited

spirits
Organic Spirits Company
The Organic Wine Company
Vinceremos Wines & Spirits

spritzer
Westons

sugar-free
Unicorn Grocery

sugars
Napier Brown & Co

supermarkets
As Nature Intended
Fresh & Wild
The Fresh Food Company
Out Of This World
Planet Organic
Sainsbury's
Simply Organic Food Company
Tesco
Waitrose

sweets
Organic Days

syrups
Napier Brown & Co

teas
Clipper Teas
Double Dragon
Equal Exchange
The Hampstead Tea and Coffee Co
Organic Days
Taylors of Harrogate
The Tea and Coffee Plant
Traidcraft plc
Wistbray

thickies
Innocent

trout
Deverill Trout Farm
Hawkshead Organic Trout Farm
Lechlade Trout Farm
Trafalgar Fisheries

turkeys
Goodman's Geese
Heal Farm Meats
Kelly Turkeys
Munson's Poultry
Northfield Farm
Woodlands Farm

vegan
Montezuma's Chocolates
Organic Oxygen

vegetables
Abel & Cole
Beano Wholefoods
The Better Food Co
Canterbury Wholefoods
Daily Bread Co-operative
Ellis Organics
Farm-A-Round
Fieldfare Organics
The Fresh Food Company
A & D Gielty
Graig Farm Organics
G's Marketing
Howbarrow Organic Farm
Marchents
Meat Matters
Myriad Organics
New Farm Organics
North East Organic Growers
Organic and Natural Foods
Organic Connections Int.
The Organic Delivery Company
The Organic Farm Shop
The Organic Shop (Online)
Organics Direct
Riverford Organic Vegetables
Scullions Organic Supplies
Sharpham Partnership
Simply Organic Food Company
South Devon Organic Producers
Sundance Market
Tamarisk Farm
Taste Connection
Thames Fruit
Total Organics
Windmill Hill City Farm Shop
Woodlands Farm

vegetarian
Bumblebee
Finlay's Foods
Goodlife Foods
Go Organic
Just Wholefoods
Mrs Moons
Naturemade
The Organic Delivery Company
Regency Mowbray Company
Tods of Orkney

vegetarian & vegan
Allergyfree Direct
Beano Wholefoods
Clearspring
Essential Trading
First Quality Foods
Get Real Organic Foods
Lyme Regis Fine Foods
On The Eighth Day Co-operative
Organics To Go
Organigo
The Pure Wine Company
Rococo Chocolates
Sundrum Organics
Vinceremos Wines & Spirits
Vintage Roots
Viva
Westons

venison
Carmichael Estate Farm Meats
Yorkshire Game

vinegar
Aspall
Crone's

water purifiers
Aqua Vitae
Fresh Water Filter Company
The Pure H20 Company
UVO (UK)

watercress
Deverill Trout Farm

wheat-free
Buxton Foods
Clearspring

wines
Bonterra Vineyards
Bumblebee
Canterbury Wholefoods
Cooks Delight
Divine Wines
ecozone.co.uk
Graig Farm Organics
Irma Fingal-Rock
Organic and Natural Foods
The Organic Delivery Company

coppice management
The 3 Ridings Coppice Group

crop protection
Tamar Organics

design for organic gardens
Les Jardiniers du Terroir

ducks
The Domestic Fowl Trust

earthworms
Wiggly Wigglers

farming
Compassion in World Farming
Elm Farm Research Centre
Federation of City Farms
Irish Organic Farmers & Growers
The Organic Advisory Service
Organic Centre for Wales
Organic Food Federation
Organic-research.com
Otley College
Rare Breeds Survival Trust
Reaseheath College
Scottish Agricultural College
Sustain
Willing Workers on Organic Farms

fertilisers
Fargro
Fertiplus Garden Products
Humberside Nursery Products Co
M S B Mastersoil Builders
Organic Concentrates
Tamar Organics

forestry
Agroforestry Research Trust
Harvest Forestry

gardening
Barwinnock Herbs
Biodynamic Agricultural Association
Country Smallholding Magazine
Dartington Tech
Defenders
Delfland Nurseries
Econopack
ecozone.co.uk
Emerson College
English Hurdle
Fargro
Ferryman Polytunnels
Fertiplus Garden Products
Green Gardener
Green Science Controls
Green Ways
Halcyon Seeds
Henry Doubleday Research Association
Horticultural Correspondence College
Humberside Nursery Products
Jekka's Herb Farm
M S B Mastersoil Builders
Mr Fothergill's Seeds
Natural Collection
Natural Surroundings
Norfolk Organic Gardeners
Organic Concentrates
Organic Food Federation
The Organic Gardening Catalogue
OrganicSeeds.co.uk
Otley College
Pershore College
Pestwatch (Bristol)
Plants for a Future
Reaseheath College
Seeds of Italy
Suffolk Herbs
Super Natural
Tamar Organics
Terre De Semences Organic Seeds
Welsh Fruit Stocks
Wiggly Wigglers
Yalding Organic Gardens

gardening information
Gardentrouble

garlic sets
Leary's Organic Seed Potatoes

geese
The Domestic Fowl Trust

herb seeds
OrganicSeeds.co.uk
Suffolk Herbs

herbs
Barwinnock Herbs
Chops Away
Dittisham Fruit Farm
Herb Society
Horticultural Correspondence College
Jekka's Herb Farm
Seeds of Italy

heritage seed library
Henry Doubleday Research Association

liquid plant food
Super Natural

livestock
Country Smallholding Magazine
Rare Breeds Survival Trust

magazine
Country Smallholding Magazine
Permaculture Magazine

mushroom spawn
Jac by the Stowl

onion sets
Leary's Organic Seed Potatoes

permaculture
Les Jardiniers du Terroir
Permaculture Association (Britain)
Plants for a Future

pest controls
Defenders
Fargro
Green Gardener
Green Science Controls
Koppert UK
Pesticide Action Network
Pestwatch (Bristol)
Stringer Laboratories

plants
Agroforestry Research Trust
Chops Away

polytunnels
Ferryman Polytunnels

potatoes
Rush Potatoes

poultry
The Domestic Fowl Trust

poultry housing
The Domestic Fowl Trust

salad seeds
Halcyon Seeds

seed potatoes
Leary's Organic Seed Potatoes

seeds
Agroforestry Research Trust
Barwinnock Herbs
Chiltern Seeds
Green Gardener
Halcyon Seeds
Humberside Nursery Products Co
Jekka's Herb Farm
Mr Fothergill's Seeds
Natural Surroundings
The Organic Gardening Catalogue
Organic Oxygen
OrganicSeeds.co.uk
Seeds of Italy
Suffolk Herbs
Tamar Organics
Terre De Semences Organic Seeds
Yalding Organic Gardens

smallholders
Country Smallholding Magazine

soft fruit plant
Welsh Fruit Stocks

soil conditioner
Super Natural

unusual seeds
Halcyon Seeds

vegan gardening
The Organic Gardening Catalogue
Plants for a Future

vegetable seeds
Chiltern Seeds
Mr Fothergill's Seeds
The Organic Gardening Catalogue
OrganicSeeds.co.uk
Seeds of Italy
Suffolk Herbs
Terre De Semences Organic Seeds

vineyard
Sedlescombe Vineyard

wildflower seeds
Suffolk Herbs

wildflowers
Jekka's Herb Farm
Natural Surroundings
The Organic Gardening Catalogue
Tamar Organics

willow products
English Hurdle

acupuncture
British Acupuncture Council
College of Integrated Chinese Medicine
London College of Traditional Acupuncture

African herbal remedies
The Little Herbal Company

Alexander technique
Soc. of Teachers of Alexander Technique

allergies
Allergyfree Direct
British Allergy Foundation
Health Flooring Network
Thorogoods (Organic Meats)

aloe vera products
The Aloe Vera Centre
Aloe Vera Health& Information
Forever Living Products
Optima Health

anthroposophic medicines
Weleda

anthroposophy
Anthroposophical Society

apitherapy
Medihoney
New Zealand Natural Food Co

aromatherapy
The Aromatic Company
Fragrant Studies Int.
H.F.M.A
Institute of Traditional Herbal Medicine & Aromatherapy
Int. Federation of Aromatherapists
Regent Academy
Scottish School of Herbal Medicine

aromatherapy products
Absolute Aromas
Active Aromatherapy
Aqua Oleum
aQuila

Aromatherapy Associates
Aromatherapy International
G Baldwin & Co
Cariad Aromatherapy
Culpeper Herbalists
Ecobrands
Eve Taylor Aromatherapy
Faith Products
Farrow & Humphreys
Fragrant Earth
Lotus Emporium
Mothernature
Natural By Nature Oils
Natures Cocoons
Naturex
Neal's Yard Remedies
NHR Organic Oils
Opal London
Ord River Tea Tree Oil Pty
Potions & Possibilities
Power Health Products
Nature's Treasures
Spice Direct
Supernature.org
Tisserand Aromatherapy
21st Century Health
Vital Touch
Women's Advisory Service

asthma
Health Flooring Network

astrology
Astrological Association (AA)

ayurveda
Ayurved Consultancy UK
Indigo Herbal

Bates method
Bates Association for Vision Education

biodegradable coffins
Greenfield Coffins

biodynamic massage
Association of Holistic Biodynamic Massage Therapists

bodycare
Active Aromatherapy
Aveda
Ayurved Consultancy UK
Bio Pathica
Blackmores
Boots Direct
Cosmetics To Go
Culpeper Herbalists
Dead Sea Spa Magik
The Deodorant Stone
Eco-Zone
ecozone.co.uk
Faith Products
Fragrant Earth
Global Eco
Grafton Int.
The Green Shop
Healthquest
Healthquest - Organic Blue
The Healthy House
Honesty Cosmetics
Kiss My Face
Lavera UK
Logona Cosmetics
M S B Mastersoil Builders
Nature's Dream
Nature's Plus
Naturex
NHR Organic Oils
Nu2trition
Opal London
Optima Health
Origins
Panacea
Pharmavita
Pitrok
Potions & Possibilities
Power Health Products
Savant Distribution
Simply Soaps
Sundance Market
Supernature.org
Think Natural
Tisserand Aromatherapy
Total Organics
21st Century Health
Veganstore
Women's Advisory Service

Bowen technique
Bowen Association. UK
Bowen Technique (Courses)
European College of Bowen Studies

brain injury
The Brainwave Centre

Buddhism
Friends of the Western Buddhist Order

cancer care
Bristol Cancer Help Centre

Chinese medicine
College of Integrated Chinese Medicine

complementary medicine
British Complementary Medicine Association
College of Naturopathic & Complementary Medicine
Complementary Medicine Association
Hale Clinic
Intecam
Natureworks
The Nutri Centre
Positive Health Magazine

complementary therapies
British School of Complementary Therapies

copper bracelets
Coppercare Products

craniosacral therapy
The Craniosacral Therapy Association

crystal therapies
Atlantis College of Crystal, Reiki & Sound Healing

dental products
Grafton Int.

deodorants
The Deodorant Stone
Pitrok

essences
Findhorn Flower Essences

essential oils
Aqua Oleum
Escential Botanicals
Ord River Tea Tree Oil Pty
The Organic Herb Trading Co

eyecare
Bates Association for Vision Education
Pure Focus Lutein Spray
Trayner Pinhole Glasses

feng shui
Feng Shui Catalogue
Feng Shui Society
Regent Academy

flotation tank
West Usk Lighthouse

flower essences
Findhorn Flower Essences

flower remedies
Dr Edward Bach Centre
Natures Cocoons

food supplements
Applejacks
G Baldwin & Co
Best Care Products
Biocare
Bio Health
Boots Direct
Country Life
Double Dragon
Earthrise
Eco-Zone
Forever Young Int.
GoodnessDirect
H.F.M.A
Healthaid
Kinesis Nutraceuticals
Maui Noni
Maximuscle
Nature's Dream
Nature's Plus
Nonu Int.
Nutricia
Nutrisport
Oceans of Goodness
Optima Healthcare

Panacea
Pharmavita
Power Products
Rapha UK
Really Healthy Company
Revital Stores
Rockland Corporation - TRC
Savant Distribution
Solgar Vitamins and Herb
Source Naturals
Thompson & Capper
Tofupill
Valuepharm
Viridian Nutrition
Well4ever
Xynergy Health Products

haircare
BiOrganic Hair Therapy
Green People
Honesty Cosmetics
Kiss My Face
Natural Science.com
Nature's Dream

headlice
Natural Science.com

heat therapy
Natural Wheat Bag Company

herbal pillows
Textures

herbal remedies
Absolutely Natural
Best Care Products
Bioforce
Bio Health
Blackmores
Body and Soul Organics
Canterbury Wholefoods
Culpeper Herbalists
Dr Edward Bach Centre
Double Dragon
Ecobrands
Fragrant earth
Fresh & Wild
H.F.M.A
Halzephron Herb Farm
Healing Herbs
Herbon
Herbs Hand Healing
Honeyrose Products
Indigo Herbal
The Little Herbal Company
Mastic Gum Europe
Mavco Health
Mycology Research Laboratories
National Institute of Medical Herbalists
Natrahealth
Naturex
Neal's Yard Remedies
Nonu Int.
On The Eighth Day Co-operative
Optima Healthcare
Panacea UK
Potter's Herbal Medicines
Pure Focus Lutein Spray
Revital Stores
Source
Spice Direct
Valuepharm
Viridian Nutrition
Weleda
Women's Advisory Service

herbal therapies
The Aloe Vera Centre
Aloe Vera Health & Information
Scottish School of Herbal Medicine
Selfheal School

herbal tonic
Green People

herbs
G Baldwin & Co
Barwinnock Herbs
Farrow & Humphreys
Halzephron Herb Farm
Herbs Hand Healing
Napiers Herbs
The Organic Herb Trading Co

holistic medicine
British Holistic Medicine Association

holistic therapy
Federation of Holistic Therapists

homoeopathic remedies
Ainsworths Homoeopathic Pharmacy
Weleda

homoeopathy
Bio Pathica
British Homoeopathic Association
British School of Homoeopathy
College of Natural Therapy
Coppercare Products
Fragrant earth
H.F.M.A
Natures Cocoons
Naturex
Neal's Yard Remedies
Revital Stores
School of Homoeopathy
Women's Advisory Service

honey
Medihoney
New Zealand Natural Food Co

hypnosis
British Society of Clinical Hypnosis
London College of Clinical Hypnosis

Indian head massage
Indian Champissage Head Massage

kinesiology
Association of Systematic Kinesiology

light therapy
Outside In (Cambridge)
S.A.D. Lightbox Co

macrobiotics
Macrobiotic Association

magazines
Bioforce
Environment & Health News
Full Moon Communications
Green Guides
Kindred Spirit
Maximuscle
Organic Living
Positive Health Magazine
Viva

massage
Association of Holistic Biodynamic Massage Therapists
The London College of Massage
Scottish School of Herbal Medicine

mastic gum
Mastic Gum Europe

medicinal herbs
Barwinnock Herbs
Cotswold Health Products
Jekka's Herb Farm

meditation
Brahma Kumaris World Spiritual Univ.
Friends of the Western Buddhist Order
The Global Retreat Centre
School of Meditation

meditation products
The Manchester Cushion Company

menopause
Tofupill

mushroom nutrition products
Mycology Research Laboratories

natural healing
Selfheal School

naturopathy
Association of Natural Medicine
The Brainwave Centre
British College of Naturopathy and Osteopathy
College of Natural Therapy
College of Naturopathic and Complementary Medicine
Gen Council & Register of Naturopaths
Passion For Life Products
Think Natural

noni juice
Maui Noni
Nonu International

nutrition
British Nutrition Foundation
College of Natural Nutrition
College of Natural Therapy
The Food Commission
Fresh Network
Greencuisine
Healthquest
Plaskett Nutritional Medicine College

nutritional supplements
Blackmores
21st Century Health

oriental medicine
London College of Traditional Acupuncture & Oriental Medicine

osteopathy
British College of Naturopathy and Osteopathy
British School of Osteopathy
General Osteopathic Council

peppermint oil
Bennett Natural Products

posture
Back in Action

pranic healing
Centre for Pranic Healing

psychotherapy
British Association of Psychotherapists

reflexology
British Reflexology Association
British School of Reflexology
Central London School of Reflexology
Int. Federation of Reflexologists
Int. Institute of Reflexology (UK)
Regent Academy

reiki
Atlantis College of Crystal, Reiki & Healing
Reiki Association

retreats
The Global Retreat Centre
Hill House Retreats: Centre
The Leela Centre
Minton House
The Retreat Association
The Retreat Company

seasonal affective disorder
Outside In (Cambridge)
S.A.D. Lightbox Co

seaweed based products
Eco-Zone
Oceans of Goodness

shiatsu
The European Shiatsu School
Ki Kai Shiatsu Centre
London College of Shiatsu
The Shiatsu College of London
Zen School of Shiatsu

skincare
Active Aromatherapy
Aromantic
Blackmores
Comfort & Joy
Dead Sea Spa Magik
Dr Hauschka Skin Care
Earthbound
Eco Clothworks
Escential Botanicals
Eve Taylor Aromatherapy
Farrow & Humphreys
Green People
Honesty Cosmetics
Honeyrose Products
Jurlique
Just Pure
Kinesis Nutraceuticals
Kiss My Face
The Little Herbal Company
Logona Cosmetics
Origins
Supernature.org
Xynergy Health Products

sound therapies
Atlantis College of Crystal, Reiki & Sound Healing

special diets
- Allergyfree Direct
- GoodnessDirect
- Potter's Herbal Medicines

special needs
- Camphill Village Trust

spiritual healing
- National Federation of Spiritual Healers

spirituality
- Brahma Kumaris World Spiritual Univ.

sports nutrition
- Applejacks
- GoodnessDirect
- Kinesis Nutraceuticals
- Mavco Health
- Maximuscle
- Nutricia
- Nutrisport

suncare
- Green People
- Kiss My Face
- Lavera UK

tea tree oil
- Health Imports
- Ord River Tea Tree Oil Pty

tobacco alternatives
- Honeyrose Products

vegan
- Green People
- Veganstore

vegetarian & vegan
- Allergyfree Direct
- Biocare
- Ethical Wares
- Hill House Retreats Centre
- Lavera UK
- Napiers Herbs

vibrational healing
- Fragrant Studies International

vitamins & minerals
- Applejacks
- Bio Health
- Boots Direct
- Kinesis Nutraceuticals
- Nature's Plus
- Nutricia
- Optima Healthcare
- Power Health Products
- Revital Stores
- Solgar Vitamins and Herb
- Thompson & Capper
- 21st Century Health
- Viridian Nutrition

water birth
- Active Birth Centre
- Association for Improvements in the Maternity Services
- The Birth Centre
- Splashdown Water Birth Services

wheat bags
- Natural Wheat Bag Company

wheatgrass products
- Earthrise
- Xynergy Health Products

yoga
- British Wheel of Yoga
- Hugger-Mugger Yoga Products
- Yoga Therapy Centre

yoga equipment
- Hugger-Mugger Yoga Products
- The Manchester Cushion Company
- Yoga Matters

activity holidays
Bicycle Beano Cycle Tours
Bike Tours
Discovery Initiatives
Long Distance Walkers Association
Wild and Free Travel
Wild Oceans Holidays
Willing Workers on Organic Farms
Woodland Skills Centre

bed & breakfasts
The Avins Bridge
Gilbert's
Higher Riscombe Farm
Home Place Farm
Howbarrow Organic Farm
Little Salkeld Watermill
Old Pines Restaurant with Rooms
J Savory Eggs
West Usk Lighthouse
Wigham Farm

café
Café@Yum
Infinity Foods
On The Eighth Day Co-operative
Pause
Pimhill Farm
The Shieling
Windmill Hill City Farm Shop

camping
Chops Away
D W & C M Evans

caravan
Peter & Therese Muskus

chalet
Peter & Therese Muskus

cottages
D W & C M Evans
Golland Farm Garden Project
Home Place Farm
Pentre Bach Holiday Cottages

cycle trailers
Living Lightly Limited

cycles
Pashley Cycles

dolphins
Wild and Free Travel

events
The Great Organic Picnic

green tourism
Sustainable Travel and Tourism
Tourism Concern

guest houses
Glenrannoch Vegetarian Guest House
Graianfryn Vegetarian Guest House
The Greenhouse Edinburgh
Lakeland Natural Veg. Guest House
Marlborough House
Yewfield Vegetarian Guest House

holidays
Argyll Hotel
The Avins Bridge Restaurant
Bicycle Beano Cycle Tours
British Trust for Conservation Volunteers
Chops Away
Cortijo Romero
Discovery Initiatives
Dittisham Fruit Farm
D W & C M Evans
Gilbert's
Glenrannoch Vegetarian Guesthouse
Golland Farm Garden Project
Graianfryn Vegetarian Guest House
The Granville Hotel
The Greenhouse Edinburgh
Higher Riscombe Farm
Home Place Farm
Hotel Mocking Bird Hill
Hugger-Mugger Yoga Products
Lakeland Natural Veg. Guest House
Lancrigg Veg. Country House Hotel
Lynford Hall Hotel
Marlborough House
Peter & Therese Muskus
Neal's Yard
Old Pines Restaurant with Rooms
Organic Holidays
Paskins Hotel
Pentre Bach Holiday Cottages
The Retreat Company
J Savory Eggs
Skyros

holidays *contd.*
Sustainable Travel and Tourism
Tamarisk Farm
Tourism Concern
Vegetarian Cookery Courses
Vegi Ventures
West Usk Lighthouse
Wigham Farm
Wild and Free Travel
Wild Oceans Holidays
Willing Workers on Organic Farms
Woodland Skills Centre
Yewfield Vegetarian Guesthouse
Youth Hostels Association

hotel
Argyll Hotel
The Granville Hotel
Hotel Mocking Bird Hill
Lancrigg Vegetarian Country House Hotel
Lynford Hall Hotel
The Organic Beef Co
Paskins Hotel
Penrhos Court Hotel and Restaurant

pub
The Duke of Cambridge

restaurant
Argyll Hotel
The Avins Bridge Restaurant
The Granville Hotel
Little Salkeld Watermill
Old Pines Restaurant with Rooms
The Organic Beef Co
Paskins Hotel
Penrhos Court Hotel and Restaurant
Sauce Organic Diner
The Village Bakery
Wild Ginger Vegetarian Bistro

retreats
The Global Retreat Centre
Hill House Retreats
The Leela Centre
Minton House
The Retreat Association
The Retreat Company

self-catering accommodation
Dittisham Fruit Farm

tipis
Peter & Therese Muskus

vegetarian
The Avins Bridge Restaurant
Bicycle Beano Cycle Tours
Glenrannoch Vegetarian Guesthouse
Lakeland Natural Veg. Guest House
Marlborough House
Paskins Hotel
Wild Ginger Vegetarian Bistro
Yewfield Vegetarian Guesthouse

vegetarian & vegan
Graianfryn Vegetarian Guest House
The Granville Hotel
The Greenhouse Edinburgh
Little Salkeld Watermill
Vegi Ventures
The Village Bakery
Wild Ginger Vegetarian Bistro

walking
Long Distance Walkers Association
The Ramblers Association

youth hostel
Youth Hostels Association

advice
- The Organic Advisory Service
- Scottish Agricultural College

books
- Ainsworths Homoeopathic Pharmacy
- Aqua Oleum
- Green Books
- Indigo Herbal
- Maya Books
- Neal's Yard Remedies
- NHR Organic Oils
- Permaculture
- Spice Direct
- Spirit of Nature
- Wiggly Wigglers

certification
- Biodynamic Agricultural Association
- Food Certification (Scotland)
- Irish Organic Farmers & Growers Association
- Soil Association

community trust
- Camphill Village Trust

conference facilities
- Minton House

consultancy
- The Organic Consultancy
- Organic Resource Agency
- Pershore College
- Reading Scientific Services
- Scottish Agricultural College

cookery courses
- Ballymaloe Cookery School
- Greencuisine
- Little Salkeld Watermill
- Mrs Tee's Wild Mushrooms
- The Natural Cookery School
- Vegetarian Cookery Courses
- The Village Bakery
- Wing of St Mawes

courses
- Alternative Therapies
- Amazon Nails Straw Bale Building
- Association of Natural Medicine
- Association of Systematic Kinesiology
- Atlantis College of Crystal, Reiki & Sound Healing
- Bowen Association UK
- Bowen Technique (Courses)
- British College of Naturopathy & Osteopathy
- British Reflexology Association
- British School of Complementary Therapies
- British School of Homoeopathy
- British School of Osteopathy
- British School of Reflexology
- Central London School of Reflexology
- Centre for Alternative Technology
- Centre for Pranic Healing
- College of Integrated Chinese Medicine
- College of Natural Nutrition
- College of Natural Therapy
- College of Naturopathic & Complementary Medicine
- Cortijo Romero
- The Craniosacral Therapy Association
- Dartington Tech
- Emerson College
- European College of Bowen Studies
- The European Shiatsu School
- Eve Taylor Aromatherapy
- Fragrant Studies International
- Horticultural Correspondence College
- Indian Champissage Head Massage
- Indigo Herbal
- Institute of Traditional Herbal Medicine & Aromatherapy
- International Institute of Reflexology
- Ki Kai Shiatsu Centre
- London College of Clinical Hypnosis
- The London College of Massage
- London College of Shiatsu
- London College of Traditional Acupuncture & Oriental Medicine
- Napiers Herbs
- Neal's Yard Remedies
- Organic Centre for Wales
- Otley College
- Permaculture Association (Britain)
- Pershore College
- Plaskett Nutritional Medicine College
- Reaseheath College
- Regent Academy

courses *contd.*

School of Homoeopathy
School of Meditation
Scottish Agricultural College
Scottish School of Herbal Medicine
Selfheal School
The Shiatsu College of London
Skyros
Walter Segal Self Build Trust
Welsh Institute of Organic Studies
Zen School of Shiatsu

directories

envirostore.co.uk
Green Books
Green Guides
Network Organic
Organic Food
The Organic Marketplace
Positive Health Magazine
Soil Association
Vegan Organic Network

education

Centre for Alternative Technology
Eco-Schools Programme
Kids Organic Club
National Association of Environmental Education
Steiner Waldorf Schools Fellowship
The Woodcraft Folk

home education

Education Otherwise
Home Education Advisory Service
Worldwide Education Service

information and learning

Centre for Alternative Technology
Organic Food

natural death

The Natural Death Centre

practitioner training

Alternative Therapies
Association of Systematic Kinesiology
British School of Complementary Therapies
British School of Homoeopathy
British School of Reflexology
Central London School of Reflexology
College of Integrated Chinese Medicine
College of Natural Nutrition
College of Natural Therapy
College of Naturopathic & Complementary Medicine
The Craniosacral Therapy Association
European College of Bowen Studies
The European Shiatsu School
Institute of Traditional Herbal Medicine & Aromatherapy
International Institute of Reflexology
Ki Kai Shiatsu Centre
The London College of Massage
London College of Shiatsu
London College of Traditional Acupuncture & Oriental Medicine
Plaskett Nutritional Medicine College
The Shiatsu College of London
Zen School of Shiatsu

publishers

Barefoot Books
Green Books
Green Guides

research

Elm Farm Research Centre
Reading Scientific Services
Scottish Agricultural College
Welsh Institute of Organic Studies

resource

envirostore.co.uk
Organic Food
Organic-research.com
Plants for a Future
Positive Health Magazine
Vegan Organic Network
Vegetarian Pages
What Doctor's Don't Tell You
WWF Global Environment Network

resource centre

Dartington Tech
The Nutri Centre

schools

Rudolf Steiner School - Kings Langley
Steiner Waldorf Schools Fellowship

stone carving course
Alternative Memorials

vegan
Vegan Organic Network
Vegan Society

vegetarian
The Natural Cookery School
Vegetarian Pages
Vegetarian Society
Vegetarian Cookery Courses

videos
Permaculture Magazine

acupuncture
British Acupuncture Council

agriculture
Biodynamic Agricultural Association

Alexander technique
Society of Teachers of Alexander Technique

allergies
British Allergy Foundation

alternative energy
British Wind Energy Association
Electric Car Association
Energy Conservation & Solar Central.
Energy Saving Trust
Solar Trade Association

alternative technology
Centre for Alternative Technology

anthroposophy
Anthroposophical Society in Great Britain

aromatherapy
International Federation of Aromatherapists

astrology
Astrological Association

beekeeping
British Beekeepers Association

Bowen technique
Bowen Association UK

building
Association of Self Builders
Association for Environment Conscious Building
Straw Bale Building Association

chiropractic
British Chiropractic Association

city farms
Federation of City Farms

complementary medicine
British Complementary Medicine Association
Complementary Medicine Association

composting
Community Composting Network

conservation
British Trust for Conservation Volunteers
Schumacher UK
Young People's Trust for the Environment & Nature Conservation

country markets
Women's Institute Country Markets

craniosacral therapy
The Craniosacral Therapy Association

death
The Natural Death Centre

economics
New Economics Foundation

education
Eco-Schools Programme
Education Otherwise
Home Education Advisory Service
Steiner Waldorf Schools Fellowship
The Woodcraft Folk

energy
National Energy Foundation

environment
Centres For Change
Earthwatch
Environ
Friends of the Earth
The Green Network
Greenpeace
National Association of Environmental Education
Soil Association
WWF Global Environment Network

ethics
Ethical Consumer
Ethical Junction

eyecare
Bates Association for Vision Education

fairtrade
The Fairtrade Foundation
Traidcraft plc

farmers' markets
National Farmers' Market Association

farming
Compassion in World Farming
Sustain
Willing Workers on Organic Farms

feng shui
Feng Shui Society

forestry
Agroforestry Research Trust
The Woodland Trust

genetics
The Genetics Forum
Genewatch UK

health foods
H.F.M.A

herbal medicine
Scottish School of Herbal Medicine

herbalism
National Institute of Medical Herbalists

herbs
Herb Society

holistic medicine
British Holistic Medicine Association

holistic therapies
Federation of Holistic Therapists

homoeopathy
British Homoeopathic Association

hostels
Youth Hostels Association

hypnosis
British Society of Clinical Hypnosis

investment
UK Social Investment Forum

macrobiotics
Macrobiotic Association of the UK

massage
Association of Holistic Biodynamic Massage Therapists

maternity
Association for Improvements in the Maternity Services
La Leche League

nappy services
National Association of Nappy Services
Real Nappy Association

natural medicine
Association of Natural Medicine
British College of Naturopathy and Osteopathy
General Council & Register of Naturopaths

nature
English Nature

nutrition
British Nutrition Foundation
Food Certification (Scotland)
The Food Commission
Fresh Network

organics
Henry Doubleday Research Association
Irish Organic Farmers & Growers
Norfolk Organic Gardeners
The Organic Advisory Service
Organic Centre for Wales
Organic Food Federation

osteopathy
British School of Osteopathy
General Osteopathic Council

permaculture
Permaculture Association (Britain)

pest control
Pesticide Action Network

psychotherapy
British Association of Psychotherapists

rare breeds
Rare Breeds Survival Trust

recycling
Community Recycling Network
National Recycling Forum
Recoup
Waste Watch

reflexology
British Reflexology Association
International Federation of Reflexologists

reiki
Reiki Association

retailing
Network of European World Shops

retreats
The Retreat Association

spiritual healing
National Federation of Spiritual Healers

tourism
Tourism Concern

vaccination
Vaccination Awareness Network

vegan
Vegan Organic Network
Vegan Society

vegetarian
Vegetarian Society

vegetarian & vegan
Viva

veterinary
British Association of Homoeopath Veterinary Surgeons

walking
Long Distance Walkers Association
The Ramblers Association

wildlife
The Wildlife Trusts

women
Women's Advisory Service
Women's Environmental Network

yoga
British Wheel of Yoga

Alphabetical List of Businesses & Organisations

Abel & Cole
www.abel-cole.co.uk

Specialities: Groceries; vegetables; fruit; meat; box service
Website: Website includes recipe page and information about organic farming. Home delivery in London and also throughout mainland UK. All produce is either bought in directly from source or can be traced back to the farm on which it was produced.
On-line ordering: Yes
Contact: Emma Hardie
Email: telesales@abel-cole.co.uk
Address: 8 - 13 MGI Estate, Milkwood Road, London SE24 0JF
Tel: 020 7737 3648
Fax: 020 7737 7785

Absolute Aromas Ltd
www.absolute-aromas.com

Speciality: Aromatherapy products
Website: Fine quality essential oils selected from assured sources around the world. The range consists of over 100 pure essential oils (cold pressed and organic wherever possible) and a variety of aromatherapy accessories. There is a product listing with prices on the website. There are also information pages on precautions and instructions are provided. Mail order is available.
Credit cards: Visa; Mastercard; Eurocard; Switch
Contact: David Tomlinson
Email: oil@absolute-aromas.com
Address: 2 Grove Park, Mill Lane, Alton, Hampshire GU34 2QG
Tel: 01420 540400
Fax: 01420 540401

Absolutely Natural Ltd
www.herbalblends.co.uk

Speciality: Herb liquid extracts
Website: Dr Benson's Herbal Blends are a range of 60 combinations of concentrated herb liquid extract. On the website there is a list of herbal blends together with an ordering facility.
On-line ordering: Yes
Credit cards: Most major credit cards
Contact: Helena Benson
Email: drbenson@herbalblends.co.uk
Address: 12 St Georges Road, London NW11 0LR
Tel: 0208 905 5509
Fax: 0208 458 9577

Active Aromatherapy
www.activearomatherapy.co.uk

Specialities: Aromatherapy products; bodycare; skincare
Website: Supplies hand-blended natural bath and body products, facial preparations and essential oils in tune with today's lifestyles. All blends are made to order. None of the ingredients have been tested on animals. Oils are listed and described and may be ordered on-line.
On-line ordering: Yes
Contact: Janet Hugill
Email: sales@activearomatherapy.co.uk
Address: Stubbings Barn, West Leith, Tring, Hertfordshire HP23 6JJ
Tel: 01442 827720
Fax: 07979 342460

Active Birth Centre
www.activebirthcentre.com

Specialities: Childbirth; birth pools; water birth
Website: A range of informative and inspirational resources covering preparation for birth and parenting, active birth and water birth, yoga pregnancy exercises, babycare, postnatal yoga and classes. There are specially selected products, including aromatherapy, for a healthy pregnancy, active birth and baby. They are described, illustrated and available for sale on-line. Birth pools are available for hire.
On-line ordering: Yes
Credit cards: Most major credit cards
Email: mail@activebirthcentre.com
Address: 25 Bickerton Road, London N19 5JT
Tel: 020 7482 5554
Fax: 020 7267 9683

AGROFORESTRY RESEARCH TRUST
www.agroforestry.co.uk

Specialities: Plants; seeds; forestry; organisation
Website: A non-profit making charity researching into temperate agroforestry. Nursery with mail order supply of plants and seeds. Also publications on agroforestry, trees, fruits, nuts and forest gardening. Website gives publications list, seed list, plant list and more details about the Trust as well as information on temperate agroforestry systems. Plants, seeds and publications may be ordered on-line as well as by phone, fax or post. Order forms are provided.
Links: Yes
On-line ordering: Yes
Credit cards: Visa; Mastercard
Email: mail@agroforestry.co.uk
Address: 46 Hunter's Moon, Dartington, Totnes, Devon TQ9 6JT
Tel: 01803 840776
Fax: 01803 840776

AINSWORTHS HOMOEOPATHIC PHARMACY
www.ainsworths.com

Specialities: Homoeopathic remedies; books
Website: Homoeopathic remedies for animals and humans. Ainsworths manufactures 3,300 remedies in a range of potencies. They also stock over 250 books on homoeopathy. There is on-line ordering for the Remedy Range, books, main range of traditional and new remedies, and remedy containers. There is an on-line magazine with a selection of relevant articles providing factual information. Mail order facilities.
Links: Yes
On-line ordering: Yes
Contact: Tony Pinkus
Email: ainshom@msn.com
Address: 38 New Cavendish Street, London W1M 7LH
Tel: 020 7935 5330
Fax: 020 7486 4313

ALARA WHOLEFOODS
www.alara.co.uk

Specialities: Nuts; dried fruits; cereals; muesli; wholesaler
Website: Manufacturers and packers of organic products. Recently launched organic muesli. Website provides details of the farms from which the products originate. It also informs on the origin, history and domestication of the plants, gives in-depth nutritional analysis, health-promoting properties and recipes. The website provides details of the product range.
Contact: Alex Smith
Email: website@alara.co.uk
Address: 110-112 Camley Street, London NW1 0PF
Tel: 020 7387 9303
Fax: 020 7388 6077

ALEMBIC PRODUCTS
www.alembicproducts.co.uk

Specialities: Food industry supplier; sauces
Website: Manufacturers of ketchup, mayonnaise, coleslaw, oil free dressings, and pasta dressings. Website provides history of the company and details of its products.
Contact: Christine Moir
Email: christine@alembicproducts.co.uk
Address: Unit 2, Brymau Estate, River Lane, Saltney, Chester, Cheshire CH4 8RQ
Tel: 01244 680147
Fax: 01244 680155

ALLERGYFREE DIRECT LTD
www.allergyfreedirect.co.uk

Specialities: Allergies; home deliveries; special diets; vegetarian & vegan
Website: Mail order specialist food products for people with allergies or dietary intolerances including gluten-free, wheat-free, dairy-free, vegan and vegetarian food. Food is GM-free and where possible organic and is delivered to the home anywhere in the UK. Catalogue and order form can be

downloaded from the website and orders may be placed by phone, fax, or email.
Links: Yes
On-line ordering: Yes
Credit cards: Most major credit cards
Email: info@allergyfreedirect.co.uk
Address: 5 Centremead, Osney Mead, Oxford, Oxfordshire OX2 0ES
Tel: 01865 722 003
Fax: 01865 244 134

The Aloe Vera Centre
www.aloeveracentre.co.uk

Specialities: Aloe vera products; herbal therapies
Website: Aloe vera is one of the oldest medicinal plants known to mankind. The website explains the properties and benefits of aloe vera and gives recommendations on products which are available through mail order but not through retail shops. There is a printable order form.
Contact: Alasdair Barcroft
Email: alasdairaloevera@aol.com
Address: PO Box 19766, London SW15 2WZ
Tel: 020 8875 9915
Fax: 020 8871 1798

Aloe Vera Health & Information Service
www.aloe.co.uk

Specialities: Aloe vera products; herbal therapies
Website: The website explains the background of the aloe vera plant and how it can help people achieve a better quality of life. The company distributes Forever Living products. These come from the largest grower, manufacturer and distributor of aloe vera in the world. Website gives information on the company and provides product information. There is a printable order form and on-line ordering is under development.
Credit cards: Visa; Mastercard; Switch
Email: sales@aloe.co.uk
Address: 55 Amity Grove, Raynes Park, London SW20 0LQ
Tel: 020 8947 6528
Fax: 020 8947 1463

Alternative Memorials
www.stonecarving.co.uk

Specialities: Memorial stones; natural death; stone-carving
Website: Alternative memorials carved with care in natural native and 'wild' stones. Stone carving is also taught on weekend courses. The website provides illustrated examples of recent commissions including headstones and memorials. Some of these are made from more unfinished stones which are not traditionally shaped. There are details of the weekend and evening stone-carving courses with dates and fees.
Contact: Will & Lottie O'Leary
Email: info@stonecarving.co.uk
Address: Upper House, Knucklas, Knighton, Powys LD7 1PN, Wales
Tel: 01547 528 792
Fax: 01547 528 792

Alternative Therapies
www.therapies.com

Specialities: Alternative therapies; courses; professional training
Website: Runs courses in reiki healing at all levels, Indian head massage, reflexology, acupressure, aromatherapy, macrobiotics, anatomy and physiology, karuna healing and oriental diagnosis. Also internet correspondence courses on anatomy, physiology and massage. Website gives details of courses and of bare-hand surgery in the Philippines. On-line facility for course booking and buying supplies and equipment.
On-line ordering: Yes
Credit cards: Visa; Mastercard; Switch
Email: info@therapies.com
Address: 16 Dukes Wood Drive, Gerrards Cross, Buckinghamshire SL9 7LR
Tel: 01753 890 202
Fax: 01753 884 069

Alternative Vehicles Technology
www.avt.uk.com

Speciality: Electric cars
Website: Electric cars, conversions and components from the leading conversion specialist in the UK. Website provides general information on electric powered vehicles and gives broad price estimates for conversions. There is also a form for joining the Electric Car Association.
Links: Yes
Email: info@avt.uk.com
Address: Blue Lias House,
Station Road, Hatch Beauchamp,
Somerset
TA3 6SQ
Tel: 01823 480 196
Fax: 01823 481 116

Amazon Nails Straw Bale Building
www.strawbalefutures.org.uk

Specialities: Straw bale buildings; courses
Website: Website offers pictures of individual straw bale building projects. There are details of building participation courses and their locations, fostering teamwork and community building. Information is given on services which include discussion of ideas, plans, preparations and materials as well as site visits.
Links: Yes
Contact: Barbara Jones
Email: barbara@strawbalefutures.org.uk
Address: Hollinroyd Farm, Butts Lane,
Todmorden, West Yorkshire OL14 8RJ
Tel: 01706 814696

Anglesey Sea Salt Co
www.seasalt.co.uk

Speciality: Salt
Website: Organic pure white natural sea salt from the seas around Anglesey. Full description on the website of the salt itself and how it is produced. There is also a spiced sea salt containing organic spices. Recipes are given and there is a printable order form for ordering direct for mail order and there is also a facility for emailing for the address of nearest retail stockist.
Credit cards: Visa; Mastercard
Contact: Kelly Lovitt
Email: enq@seasalt.co.uk
Address: Llanfair PG, Anglesey,
Gwynedd LL61 6TQ, Wales
Tel: 01248 430871
Fax: 01248 430213

Animal Friends Insurance
www.animalfriends.org.uk

Speciality: Pet insurance
Website: Offers a truly ethical insurance alternative. All net profits are dedicated to caring for animals. Home, pet, motor, travel and life insurance are offered. On-line quotation facilities are available on the website as well as an ordering facility. The website gives celebrity endorsements.
On-line ordering: Yes
Contact: Christopher Fairfax
Email: info@animalfriends.org.uk
Address: Suite 4M,
Eastmill, Belper,
Derbyshire DE56 2UA
Tel: 0870 444 3438

Anthroposophical Society in Great Britain
www.anth.org.uk/rsh

Specialities: Anthroposophy; organisation
Website: The home of anthroposophy in the UK. The site gives information on lectures, classes, group work, weekend workshops, events, exhibitions, house facilities and performances.
Email: RSH@cix.compulink.co.uk
Address: Rudolph Steiner House,
35 Park Road, London NW1 6XT
Tel: 020 7723 4400
Fax: 020 7724 4364

Applejacks
www.applejacks.co.uk

Specialities: Food supplements; sports supplements; vitamins
Website: Company provides choices for health, fitness, weight loss, beauty, women's health, diets, skin, hair and body care, body-building, sports fitness and training. Products are catalogued on-line with descriptions and prices.
On-line ordering: Yes
Credit cards: Visa; Mastercard; Switch
Email: robert@applejacks.co.uk
Address: 28 The Mall, The Stratford Centre, London E15 1XD
Tel: 020 8519 5809
Fax: 020 8519 1099

Aqua Oleum
www.aqua-oleum.co.uk

Specialities: Aromatherapy products; books
Website: Many years experience in sourcing and supplying essential oils, dealing directly with the place of origin and production of each oil. Most of the comprehensive range is produced from organically grown or wild plants. This small family business offers an on-line ordering facility. Also available are books written by Julia Lawless.
On-line ordering: Yes
Credit cards: Visa; Mastercard
Contact: Alec & Julia Lawless
Email: info@aqua-oleum.co.uk
Address: Unit 3, Lower Wharf, Wallbridge, Stroud, Gloucestershire GL5 3JA
Tel: 01453 753555
Fax: 01453 752179

Aqua Vitae
www.aquavitae.co.uk

Speciality: Water purifiers
Website: Water purifiers with micro-porous ceramic filters which provide fresh and clean water on tap whenever required. The website explains the principles of the purification process, illustrates the range, gives prices and provides an on-line ordering facility. There is a list of events at which the company exhibits.
On-line ordering: Yes
Credit cards: Most major credit cards
Email: info@aquavitae.co.uk
Address: Broadhembury, Honiton, Devon EX14 3NJ
Tel: 01404 841 841
Fax: 01404 841 610

aQuila
www.aquiladirect.com

Speciality: Aromatherapy products
Website: Specialises in hand blended natural aromatherapy skin products which are vegetable, plant, fruit, herb mineral or vitamin based and which have not been tested on animals. Website headings cover personal prescriptions, sensuous products, female face and body, male face and body, children, therapeutic, and DIY aromatherapy. Orders can be placed on-line or over the phone with payment by cheque.
On-line ordering: Yes
Credit cards: Visa; Mastercard
Address: 8 Stockdale Close, Knaresborough, North Yorkshire HG5 8EA
Tel: 01423 861 208

ARCHIAMMA
www.archiamma.co.uk

Specialities: Curry kits; spices
Website: Small family-run business roasting and grinding organic spices for Sri Lankan dishes. The kits include all spices as well as recipe information. Kits are prepared to order. There is a printable order form and the site also provides information on fairtrade, organic farming in the developing world and the forest garden system in Sri Lanka.
Contact: Nelisha Wickremasinghe
Email: nelisha@archiamma.co.uk
Address: 1/2 Upper Dully Cottages, Dully Road, Sittingbourne, Kent ME9 9PA
Tel: 01795 520504
Fax: 01795 520409

Architectural Salvage Index
www.handr.co.uk

Speciality: Building materials
Website: A register of every type of recyclable building material available in the UK. Architectural salvage is part of the larger Hutton & Rostron's website. There is an explanation of how the site works, how to register items for sale and how to consult items available. A list of dealers in a given area may be ordered.
Email: admin@handr.co.uk
Address: Hutton & Rostron, Netley House, Gomshall, Guildford, Surrey GU5 9QA
Tel: 01483 203221
Fax: 01483 202911

Architype
www.architype.co.uk

Speciality: Architects
Website: Ecological and sustainable architecture and building design. Website provides details of different project types from self-build to ecological. It includes a show house designed and built at the Centre of Alternative Technology.
Contact: Jon Broome
Email: jon@architype.co.uk
Address: 1b Leather Market Street, London, SE1 3JA
Tel: 020 7403 2889
Fax: 020 7407 5283

Argyll Hotel
www.argyllhoteliona.co.uk

Specialities: Hotel; restaurant; holidays
Website: Small hotel serving own-grown organic foods and local organic meat. Website sets out to be an annotated database for the island, the hotel and the surrounding area. Email booking form is provided. There is an aerial map and comprehensive instructions on how to get to Iona.
Contact: Claire Bachellerie
Email: reception@argyllhoteliona.co.uk
Address: Isle of Iona, Argyll & Bute PA76 6SJ, Scotland
Tel: 01681 700 334
Fax: 01681 700 510

Aromantic
www.aromantic.co.uk

Speciality: Skincare
Website: Courses in making one's own beauty products The raw materials for skincare and beauty products can be ordered from the website. There is a full on-line catalogue of the range. Information on training courses is provided and shortly there will be an on-line booking facility for courses.
Links: Yes
On-line ordering: Yes
Credit cards: Most major cards
Email: info@aromantic.co.uk
Address: 4 Heathneuk, Findhorn, Moray IV36 3YY, Scotland
Tel: 01309 692000
Fax: 01309 691100

Aromatherapy Associates
www.aromatherapyassociates.com

Speciality: Aromatherapy products
Website: The on-line catalogue is divided into summer specials, body, mind, spirit, skin care, home, travel and gifts. There is an on-line ordering facility and a faxable order form. A la carte treatments are available in London and details are given on the site together with endorsements.
On-line ordering: Yes
Credit cards: Most major credit cards
Export: Yes
Email:
info@aromatherapyassociates.com
Address: 68 Maltings Place, Fulham, London SW6 2BY
Tel: 020 7371 9878
Fax: 020 7371 9894

Aromatherapy International Ltd
www.aromainter.com

Speciality: Aromatherapy products
Website: The company represents farmers who are producing organic essential oils in several countries. This means employment for the indigenous population and a contribution to helping protect the environment. A range of over 50 certificated organic essential oils and over 20 organic vegetable oils are available. The range includes new organic oils from Madagascar. Website was under construction at time of writing.
Contact: John Brebner
Email: sales@aromainter.com
Address: Banks House, Baileys Lane, Burton Overy, Leicestershire LE8 9DD
Tel: 0116 259 0055
Fax: 0116 259 3033

The Aromatic Company
www.aromatic.co.uk

Speciality: Aromatherapy products
Website: The Aromatic Company web shop is divided into four section headings on the website: remedies, essential oils and aromatherapy cases, the animal range and the tattoo range. On-line ordering is available from each section. Aromatherapist advice on ailments is available free via email.
Links: Yes
On-line ordering: Yes
Export: Yes
Email: sales@aromatic.co.uk
Address: LINCS Aromatics Ltd, Unit 18, Thames Street, Louth, Lincolnshire LN11 7AD
Tel: 0800 9961148

As Nature Intended
www.asnatureintended.uk.com

Speciality: Supermarket
Website: High street retailer of organic foods, vitamins and lifestyle products. Website gives details of the company, its philosophy, product range, property and career opportunities. The company has an ambitious growth programme.
Contact: Pippa Sterry
Email: enquiries@asnatureintended.uk.com
Address: Unit 3, Roslin Square, South Acton, London W3 8DH
Tel: 020 8752 0468
Fax: 020 8752 0418

Ashfield Organic Farm
www.ashfieldfarm.8m.com

Specialities: Farm shop; box scheme
Website: Website gives vegetable price list. Site was under revision at the time of writing and was due to be enlarged.
Links: Yes
Contact: Geoff Nicholls
Email: enquiries@ashfieldfarm.co.uk
Address: Chester High Road, Neston, Cheshire CH64 3RY
Tel: 0151 336 8788

Aspall
www.aspall.co.uk

Specialities: Apple juice; vinegars; cyder
Website: Products are available through supermarkets and health food shops. Juice is made from whole fruits and not from concentrates. Aspall is the UK's sole producer of organic cyder vinegar. The website gives extensive information on the history, manufacture and uses of vinegar, as well as suggested recipes. There is an email enquiry facility.
Contact: Barry Chevallier-Guild
Email: info@aspall.co.uk
Address: The Cyder House, Aspall Hall, Stowmarket, Suffolk IP14 6PD
Tel: 01728 860510
Fax: 01728 861031

Association for Environment Conscious Building
www.aecb.net

Specialities: Building advice; organisation
Website: Advice on all aspects of 'green building'. It produces in book form a directory of environmentally friendly building services and products. The website gives information and news and there is also an on-line search facility for members in given areas. There is a book service and a membership application form which may be printed from the site. The charter of the association is available on-line.
Email: admin@aecb.net
Address: Nant-Y-Garreg, Saron, Llandysul, Carmarthenshire SA44 5EJ, Wales
Tel: 01559 370908
Fax: 01559 370908

Association for Improvements in the Maternity Services
www.aims.org.uk

Specialities: Birth choices; water births; organisation
Website: The association exists to give help and information on birth choices, especially for women who wish to have home and water births and who are seeking information on ante-natal tests such as ultrasound. The website provides news and articles, publications listing and joining information.
Links: Yes
Contact: Beverley Lawrence Beech
Email: chair@aims.org.uk
Address: 21 Iver Lane, Iver, Buckinghamshire SL0 9LH
Tel: 01753 652781
Fax: 01753 654142

Association of Holistic Biodynamic Massage Therapists
www.ahbmt.org

Specialities: Massage; bio-dynamic massage; organisation
Website: A professional organisation that seeks to promote bio-dynamic massage and encourage high standards in its practice. The website gives an explanation of bio-dynamic massage and information on training establishments. There is an on-line form for requesting further information including finding local practitioners.
Contact: K Stauffer
Email: kathrin@ahbmt.org
Address: 42 Catherine Street, Cambridge, Cambridgeshire CP1 3AW
Tel: 01223 240 815

Association of Natural Medicine
www.anm.org.uk

Specialities: Natural medicine; organisation; courses
Website: The website provides information on the association, its aims, history and code of ethics. It also gives details of diploma courses available in anatomy and physiology, acupuncture, aromatherapy, counselling, homoeopathy, sports injury massage, therapeutic massage, hypnotherapy, diet and nutrition, reflexology, kinesiology and non-manipulative re-alignment therapy.
Email: association.naturalmedicine@talk21.com
Address: 19a Collingwood Road, Witham, Essex CM8 2DY
Tel: 01376 502 762
Fax: 01376 502 762

Association of Self Builders
www.self-builder.org.uk

Specialities: Self-build; organisation
Website: Provides the opportunity of meeting other self-builders who can offer advice and experience. The website gives information on the benefits of the association, members' builds, and meetings. There is an on-line joining form. This is a voluntary organisation. Site visits can be arranged.
Links: Yes
Email: fadamson@paulassociates.com
Address: Room 23, The Rufus Centre, Steppingley, Flitwick, Bedfordshire MK45 1AN
Tel: 0704 154 4126
Fax: 0704 154 4126

Association of Systematic Kinesiology
www.kinesiology.co.uk

Specialities: Kinesiology; courses; practitioner training
Website: Courses in muscle testing, analysis and holistic balancing. Courses include a foundation course and a two year diploma course, details of which are given on the website. There is a list of books and tapes for those practising authentic kinesiological muscle testing. The Association will advise on the nearest practitioner to a given location.
Email: info@kinesiology.co.uk
Address: 39 Brown's Road, Surbiton, Surrey KT5 8ST
Tel: 020 8399 3215
Fax: 020 8390 1010

Astrological Association (AA)
www.astrologer.com/aanet

Specialities: Astrology; organisation
Website: The main purpose is to bring astrology out of the fringe into society's mainstream. It currently has about 1,000 members in the UK. The website gives information on the association, membership, joining.and astrology in higher education.
Links: Yes
Email: astrological.association@zetnet.co.uk
Address: Unit 168, Lee Valley Techno Park, Tottenham Hale, London N17 9LN
Tel: 0906 716 003
Fax: 020 8880 4849

Atlantis College of Crystal, Reiki and Sound Healing
www.atlantiscrystalhealing.com

Specialities: Crystal therapies; reiki; sound therapies; courses
Website: The college specialises in a wide range of complementary therapy and teaches reiki, seichem, sound, meditation, crystal healing, languages of light and chakra healing to master and teacher levels. Website gives details of correspondence courses.
Links: Yes
Contact: Lesley Carol
Email: enquiries@atlantiscrystalhealing.com
Address: Higher Washford Farm, Washford, Watchet, Somerset TA23 0NS
Tel: 01984 640588
Fax: 01984 640588

Auro Organic Paint Supplies
www.auroorganic.co.uk

Specialities: Paints
Website: Importers of natural organic paints including emulsions, eggshells, glosses, floor finishes, woodstains, adhesives and varnishes. No petro-chemicals are used in their formulations. Website gives full listing and description of products and their uses. There is an instructive page on allergies. There is an email form for specific enquiries.
Contact: Anne Westbrook
Email: sales@auroorganic.co.uk
Address: Unit 2, Pamphillions Farm, Debden, Saffron Walden, Essex CB11 3JT
Tel: 01799 543 077
Fax: 01799 542 187

Aveda
www.aveda.com

Speciality: Bodycare
Website: An international company with a very extensive range of health and beauty products. There is a full product listing on the website which also has wide-ranging information on the company.
Address: 28 Marylebone High Street, London W1U 4PL
Tel: 020 7224 3157

The Avins Bridge Restaurant & Rooms
www.theavinsbridge.co.uk

Specialities: Restaurant; bed & breakfast; holidays; vegetarian
Website: Restaurant specialising in organic food and wine with bed and breakfast accommodation. The website provides a sample menu and details of the accommodation and prices. There is an on-line email booking form.
Links: Yes
Email: enquiries@theavinsbridge.co.uk
Address: College Road, Ardingly, Haywards Heath, West Sussex RH17 6SH
Tel: 01444 892393

Ayurved Consultancy UK
www.ayurved-herbeli.com

Specialities: Bodycare: ayurveda
Website: Indian herbal medicine for health and well-being. Website gives brief introduction to ayurveda, information on Dr Ela Shah, and offers an on-line shop for bodycare creams.
Links: Yes
On-line ordering: Yes
Catalogue: Yes
Contact: Dr Ela Shah
Email: info@ayurved-herbeli.com
Address: 50 Elmcroft Crescent, London NW11 9SY
Tel: 020 8455 6598
Fax: 020 8455 6598

Baby Organix (Organix Brands)
www.babyorganix.co.uk

Speciality: Baby food
Website: Website has been created to give detailed advice and to answer the most frequently asked questions about weaning a baby. It also gives details on making babyfood recipes at home including full nutritional information. There are details of stockists and where to buy the food on-line.
Address: Knapp Mill, Mill Road, Christchurch, Dorset BH23 2LU
Tel: 0800 393 511
Fax: 01202 479 712

Back in Action
www.backinaction.co.uk

Specialities: Chairs; back trouble; posture
Website: The company was formed to help prevent and alleviate back pain through healthy furniture and posture. Website provides a taster of some of the chairs and gives details of showrooms where the range can be viewed. There are chairs for the office and special chairs for children.
Catalogue: Information pack
Email: info@backinaction.co.uk
Address: Back In Action, 11 Whitcomb Street, London WC2H 7HA
Tel: 020 7930 8308
Fax: 020 7925 0250

G Baldwin & Co
www.baldwins.co.uk

Specialities: Herbs; aromatherapy products; food supplements
Website: Secure on-line shopping for dried herbs, aromatherapy oils, tinctures, fluid extracts, nutritional supplements, incense, ingredients for cosmetic making and books.
Links: Yes
On-line ordering: Yes

cards: Visa; Mastercard; Switch; JCB; Amex
Catalogue: Yes
Email: sales@baldwin.co.uk
Address: 171-173 Walworth Road, London, SE17 1RW
Tel: 020 7703 5550
Fax: 020 7252 6264

Ballymaloe Cookery School
www.ballymaloe-cookery-school.ie

Speciality: Cookery school
Website: Promotes the unique style of cooking pioneered by Myrtle Allen at Ballymaloe House. Most of the vegetables and herbs used are grown organically in the adjoining gardens and greenhouses. Free range eggs come from the school's own hens and meat and cheeses are provided by the best local producers. Website gives details of cookery courses and provides an on-line booking form. There is information on accommodation as well as on lifestyle and gardening courses.
Contact: Darina & Tim Allen
Email:
info@cookingisfun.ie
Address: Shanagarry, Midleton, Co Cork, Ireland
Tel: + 353 (0) 21 4646 785
Fax: + 353 (0) 21 4646 909

Bambino Mio
www.bambino.co.uk

Speciality: Nappies
Website: Website provides information on the company and why one might choose cotton nappies. There is a full product range described on the site from cotton nappies, to nappy covers, training pants, swim nappies, nappy bags and nappy washing powder. There is a price list and on-line ordering. There is also a facility for locating local stockists.
On-line ordering: Yes
Credit cards: Most major credit cards
Catalogue: Free brochure available
Email: enquiries@bambino.co.uk
Address: 12 Staveley Way, Brixworth, Northampton NN6 9EU
Tel: 01604 883 777
Fax: 01604 883 666

Barefoot Books Ltd
www.barefoot-books.com

Specialities: Publisher; children's books
Website: Independent children's book publisher which celebrates art and story with books that open the hearts and minds of children from all walks of life, inspiring them to read deeper, search further and explore their own creative gifts. Website has section headings for family hearth, teachers' tent, artists' café and storytellers' caravan. There is a browse-and-buy facility.
On-line ordering: Yes
Catalogue: Yes
Email: info@barefoot-books.com
Address: 18 Highbury Terrace, London N5 1UP
Tel: 020 7704 6492
Fax: 020 7359 5798

Bart Spices Ltd
www.bartspices.com

Specialities: Herbs; spices
Website: Producers of organic herbs and spices which are sold through the retail trade and to other manufacturers. The website is a source of information on herbs and spices and a recipe archive. There is a regional recipe index as well as a spice index.
Email: bartspices@bartspices.com
Address: York Road, Bristol BS3 4AD
Tel: 0117 977 3474
Fax: 0117 972 0216

Barwinnock Herbs
www.barwinnock.com

Specialities: Herbs; medicinal herbs; culinary herbs; fragrant leaved plants; seeds; gardening
Website: Barwinnock uses organic gardening principles to grow herbs, medicinal plants, fragrant leaved plants and wildflowers. Site headings include a useful herb finder by Latin or common name, bog plants and native wildflowers. There is advice on perfumed gardens, fragrant hedges, chamomile lawns and comfrey root cuttings. A seed list with prices is also provided and an order form can be printed from the website.
On-line ordering: No
Catalogue: Yes (3 - 1st class stamps)
Contact: Dave & Mon Holtom
Email: herbs.scotland@barwinnock.com
Address: Barrhill, by Girvan, Ayrshire KA26 0RB, Scotland
Tel: 01465 821338
Fax: 01465 821338

Bates Association for Vision Education (BAVE)
www.seeing.org

Specialities: Eyesight; Bates method; organisation
Website: The association teaches people how to improve their eyesight without lenses or surgery and how to increase their chances of healing in cases of degenerative eye disease. Website provides an introduction to the history, techniques and books. It also has case histories from around the world, the latest news, finding a teacher, advertisements and many links.
Links: Yes
Email: bave@seeing.org.net
Address: PO Box 25, Shoreham by Sea, West Sussex BN43 62F
Tel: 01273 422 090
Fax: 01273 279 983

Beaming Baby.com
www.beamingbaby.com

Specialities: Babycare; nappies
Website: On-line shopping for everything natural and organic for mother, baby and toddler. The baby section is divided into bath, body, bottom, health, travel and tummy and the mother section into body, mother, home essentials, family health, family travel and family aromas. There is an email request form for a printed catalogue.
On-line ordering: Yes
Credit cards: Most major credit cards
Catalogue: Yes
Email: service@beamingbaby.com
Address: 25 Ellerhayes, Hale, Exeter, Devon EX5 4PU
Tel: 0800 0345 672
Fax: 01392 882 332

Beano Wholefoods
www.beanowholefoods.co.uk

Specialities: Fruit; vegetables; home delivery; vegetarian & vegan
Website: Stocks wide range of organic food. Delivery service available in the Leeds area. The history of the workers co-operative is given on the website. There is a product listing, directions and a map.
Email: info@beanowholefoods.co.uk
Address: 36 New Briggate, Leeds, West Yorkshire LS1 6NU
Tel: 0113 243 5737
Fax: 0113 243 5737

The Beer Shop
www.pitfieldbeershop.co.uk

Specialities: Beers; brewing equipment
Website: Mail order service. All Pitfield beers are now certified organic, both bottled and cask. There is a full listing of beers available. The site is small and relatively basic.
Credit cards: All major cards
Email: sales@pitfieldbeershop.co.uk
Address: 14 Pitfield Street, London N1
Tel: 020 7739 3701

Belvoir Fruit Farms
www.belvoircordials.co.uk

Speciality: Cordials
Website: Organic sparkling elderflower pressé, ginger beer, lemonade, and blackcurrant, ginger and lemon cordials. Cordials are produced using only fresh fruits, flowers and spices with no artificial colourings or sweeteners. A full range of products is described on the site and recipes are given. There is also a company profile. Major stockists are listed and information on regional and local stockists may be requested.
Links: Yes
Contact: Peverell Manners
Email: sales@belvoircordials.co.uk
Address: Belvoir, Grantham, Lincolnshire NG 32 1PB
Tel: 01476 870286
Fax: 01476 870114

Bennett Natural Products
www.healthremedies.co.uk

Speciality: Peppermint
Website: The leading brand of oil of peppermint for many years. Manufactured in Denmark, there are tablets, capsules, powder and pure Japanese oil of peppermint. Website gives contact information only.
Contact: Mark Bennett
Email: jmb.bennett@sagehost.co.uk
Address: Enterprise House, Richmond Hill, Pemberton, Wigan, Greater Manchester WN5 8DT
Tel: 01257 404659
Fax: 01257 404671

Best Care Products
www.bestcare-uk.com

Specialities: Food supplements; herbal remedies; detoxification
Website: Offers the Robert Gray Intestinal Cleansing programme as well as various teas, minerals and baths salts to detoxify and cleanse the body. Full instructions and explanations are provided on the website. There is a product listing and email order form.
Credit cards: Most major cards
Contact: Hermann Keppler
Email: info@bestcare-uk.com
Address: 73 Gardenwood Road, East Grinstead, West Sussex RH19 1RX
Tel: 01342 410303
Fax: 01342 410909

Better Baby Sling
www.betterbabysling.co.uk

Speciality: Baby carriers
Website: An over-the-shoulder carrier suitable for single babies and twins. The sling is designed to hold babies close to the body in the most natural way. The website explains the principles of the sling and gives instructions on using it. Slings are available in a number of different fabrics. There is an instructional video showing a range of positions for the use of the slings.
Links: Yes
On-line ordering: Yes
Credit cards: Visa; Mastercard; Delta; Switch; Solo.
Contact: Sue Budden
Email: info@betterbabysling.co.uk
Address: 47 Brighton Road, Watford, Hertfordshire WD24 5HN
Tel: 01923 444 442
Fax: 01923 444 440

The Better Food Co
www.betterfood.co.uk

Specialities: Meat; vegetables; fruit; groceries; household products; box scheme
Website: Company provides a personal home delivery service and also supplies on a wholesale basis. Site was under construction at time of writing.
Contact: Phil Houghton
Email: betterfood@compuserve.com
Address: Unit 1, Wallis Estate, Mina Road, Bristol BS2 9YW
Tel: 0117 904 1191
Fax: 0117 904 1190

Bicycle Beano Cycle Tours
www.bicycle-beano.co.uk

Specialities: Activity holidays; holidays; vegetarian
Website: Sociable cycling holidays on country lanes in Wales and the Welsh borders with an emphasis on vegetarian cuisine, relaxation and enjoyment. Website provides an introduction to the tours, details of bicycling weeks and weekends, essential information and reservation information.
Email: mail@bicycle-beano.co.uk
Address: Erwood, Builth Wells, Powys LD2 3PQ, Wales
Tel: 01982 560 471

BigBarn
www.bigbarn.co.uk

Speciality: Local produce nationwide
Website: Interactive website putting fresh produce on the map. The company has access to over 3,000 independent food producers. Simply enter your town or postcode in the search facility and a map will appear with local producers coded by icon with contact details. There is also direct access to the websites of a number of member producers. There is on-line registration for regular special offers from local suppliers.
Links: Yes
Credit cards: Most major credit cards
Email: postmaster@bigbarn.co.uk
Address: College Farm, High Street, Great Barford, Bedford MK44 3JJ
Tel: 01234 871 580

Big Oz Industries Ltd
www.bigoz.co.uk

Specialities: Cereals; Australian cereals; wholesalers
Website: Importer of organic breakfast cereals from Australia. Website gives information on individual breakfast cereals and their ingredients, nutritional information and special diets. There are details of mail order selection boxes which can be ordered on-line.
On-line ordering: Yes
Contact: Anne Lotter
Email: anne@bigoz.co.uk
Address: PO Box 48, Twickenham, Middlesex TW1 2UF
Tel: 020 8893 9366
Fax: 020 8894 3297

Bike Tours Ltd
www.biketours.co.uk

Specialities: Activity holidays; holidays
Website: Exciting and diverse cycling holidays throughout Europe. Luggage is transported separately for the cyclist. Tours are graded according to terrain by entry level, intermediate and experienced. The tours are all listed on the website but a printed brochure is also available. Bookings can be made on-line.
On-line ordering: Yes
Credit cards: Most major credit cards
Catalogue: Brochure
Email: mail@biketours.co.uk
Address: Victoria Works, Lambridge Mews, Larkall, Bath BA1 6QE
Tel: 01225 310 859
Fax: 01225 480 132

Bill Dunster Architects
www.zedfactory.com

Speciality: Architects
Website: Operates on the belief that sustainable development is both affordable and possible within current market constraints. Their expertise lies in low energy as well as in low environmental impact building and landscape. Projects are listed on the site and the schemes can be viewed in detail.
Email: bill.dunster@btinternet.com
Address: Hope House, Molember Road, East Molesey, Surrey KT8 9NH
Tel: 020 8339 1242
Fax: 020 8339 0429

The Billington Food Group
www.billingtons.co.uk

Specialities: Cane sugars; food industry supplier
Website: Privately owned family company which supplies a range of cane organic cane sugars to retailers and manufacturers. Sugars are unrefined and simply processed to retain the cane's natural molasses. Website gives company details, product range and stockist information. There is also information on organic sugar.
Email: bfg@billingtons.co.uk
Address: Cunard Building, Liverpool, Merseyside L3 1EL
Tel: 0151 236 2265
Fax: 0151 236 2493

Biocare Ltd
www.biocare.co.uk

Specialities: Food supplements; vegetarian & vegan
Website: The company's range embraces everything from vitamins to probiotics and from herbal extracts to neutriceutical combinations. In doing so it caters for vegetarians, vegans and those wishing to avoid animal-based products. Website under construction at time of writing but will contain information on products, articles on natural healthcare and contact facilities.
Contact: Robert Joy
Email: rcjoy@btconnect.com
Address: The Lakeside Centre, 180 Lifford Lane, Kings Norton, Birmingham, West Midlands B30 3NU
Tel: 0121 433 8711
Fax: 0121 433 8707

Biodynamic Agricultural Association
www.anth.org.uk/biodynamic

Specialities: Organisation; bio-dynamic cultivation; certification; agriculture; gardening
Website: Part of a worldwide movement promoting the bio-dynamic approach to organic agriculture which developed from Rudolf Steiner's spiritual scientific research. It offers training in bio-dynamic farming and gardening and also certifies produce grown bio-dynamically. Website sets out aims and objectives and provides information on work opportunities, training, conferences, bio-dynamic supplies, seed work and the library of articles.
Links: Yes
Email: bdaa@biodynamic.freeserve.co.uk
Address: Painswick Inn Project, Gloucester Street, Stroud, Gloucestershire GL5 1QG
Tel: 01453 759501
Fax: 01453 759501

Bioforce
www.bioforce.co.uk

Specialities: Herbal remedies; magazine
Website: Bioforce, the fresh herb company is a modern science-based organisation grounded in the knowledge and traditions of herbalists and naturopaths. Herbs, organically grown from genetically pure seeds, are processed whilst still fresh to good manufacturing practice standards. Publishers of *Health Way* magazine. Extensive and informative website with pages on echinaforce, phytotherapy; healthfoods, bodycare and the magazine as well as details on the company.
Email: enquiries@bioforce.co.uk
Address: Bioforce (UK) Ltd, 2 Brewster Place, Irvine, Ayrshire KA11 5DD, Scotland
Tel: 01294 277344
Fax: 01294 277922

Bio Health Ltd
www.bio-health.ltd.uk

Specialities: Herbal remedies; food supplements; vitamins & minerals
Website: Independent manufacturers and suppliers of a unique range of top quality, additive-free products. Company produces quality supplements, promotes natural healthcare products and provides expert advice and information. The website gives an indexed listing of products available with capacities and prices.
Contact: June Crisp
Email: info@bio-health.ltd.uk
Address: Culpepper Close, Medway City Estate, Rochester, Kent ME2 4HU
Tel: 01634 290115
Fax: 01634 290761

Bio Pathica
www.biopathica.com

Specialities: Functional foods; homoeopathy; bodycare
Website: Functional food (spirulina-chlorella), probiotics, heel antihomotoxic medicines, heomeoden/heel homoeopathics, natural skin care, homoeopathic products shown on the web pages can only be obtained through professional healthcare practitioners, pharmacies and on prescription. An extensive website with product listings and a wide range of information.
Contact: Roger Wilson
Email: products@biopathica.com
Address: PO Box 217, Ashford, Kent TN23 6ZU
Tel: 01233 636678
Fax: 01233 638380

BiOrganic Hair Therapy
www.biorganics.co.uk

Speciality: Haircare
Website: Organic hair products made for hairdressers but also available direct by mail order. There is an order form with prices to print out. Website gives information on cleansing, conditioning, styling, weight loss, deodorising and cruelty-free beauty.
Email: enquiry@biorganics.com
Address: Unit 2 Caxton Park, Wright Street, Old Trafford, Manchester M16 9EW
Tel: 0161 872 9813
Fax: 0161 872 9848

The Birth Centre
www.birthcentre.com

Specialities: Childbirth; water birth
Website: The birth centre provides a special environment for giving birth with rooms with birthing pools and garden views. The website provides a full explanation of the facilities, birth information, details on the midwives and facts and figures. A free consultation is available.
Links: Yes
Catalogue: Free brochure
Email: midwifecf@aol.com
Address:
Caroline Flint Midwifery Services, 34 Elm Quay Court, Nine Elms Lane, London SW8 5DE
Tel: 020 7498 2322
Fax: 020 7498 0698

Blackmores Ltd
www.blackmores.com

Specialities: Skin and haircare products; herbal remedies; nutritional supplements
Website: An international website with headings on health, lifestages, healthy lifestyle, health news, on-line naturopath and on-line newsletters The company is Australian based and products are available in the UK through health shops and also by mail order.
Email: blackmoresuk@intonet.co.uk
Address: Willow Tree Marina, West Quay Drive, Yeading, Middlesex UB4 5TA
Tel: 020 8842 3956
Fax: 020 8841 7557

Blackwall Ltd
www.blackwall-ltd.com

Specialities: Composting; recycling
Website: Specialists in environmental product promotions and mail order. Products include gardening, water efficiency, home composting and kerbside recycling. Website shows and describes products many of which are available by mail order direct to the public.
Catalogue: Yes
Email: info@blackwall-ltd.com
Address: Seacroft Estate, Coal Road, Leeds, West Yorkshire LS14 2AQ
Tel: 0113 201 8000
Fax: 0113 201 8001

Blendex Food Ingredients
www.blendex.co.uk

Specialities: Blenders of organic herbs and spices; food industry supplier
Website: Blenders of organic herbs and spices for the food industry. There is a listing of products on the website as well as background company information.
Contact: N J Robinson
Email: blendex@blendex.co.uk
Address: Hetton Lyons Industrial Estate, Hetton le Hole, Tyne and Wear DH5 0RG
Tel: 0191 517 0944
Fax: 0191 526 9546

Body and Soul Organics
www.organic-gmfree.co.uk

Specialities: Vegetables; fruit; meat; herbal remedies; gluten-free; dairy-free
Website: Supplier of organic food free of GMOs. Home delivery service throughout the UK from this retail outlet which carries over 2,000 organic lines including fruit, vegetables, eco-friendly household products, herbal remedies and toiletries and support for people on special diets. A complete catalogue can be downloaded from the website. Orders can be emailed, phoned or faxed.
Email: bodyandsoul@organic-gmfree.co.uk
Address: 1 Parade Court, Ockham Road South, East Horsley, Surrey KT24 6QR
Tel: 01483 282868
Fax: 01483 282060

Bodywise (UK) Ltd
www.natracare.co.uk

Specialities: Tampons; sanitary products; babycare
Website: 100% cotton tampons with and without applicators. Guaranteed GMO-free and certified organic. Mail order available. Website gives information on feminine hygiene, organic tampons, health issues and babycare. It also advises on where to buy its products and provides links to mail order and internet suppliers.
Links: Yes
Contact: Susie Hewson-Lowe
Email: sales@natracare.co.uk
Address: Unit 23, Marsh Lane Industrial Estate, Marsh Lane, Portbury, Gloucestershire BS20 0NH
Tel: 01275 371764
Fax: 01275 371765

Bonterra Vineyards
www.bonterra.com

Speciality: Wine
Website: Californian vineyard which prides itself on producing organic wines using no herbicides, pesticides, fungicides, artificial fertilisers or synthetic chemicals. The emphasis is placed on building a living soil and environment that encourages beneficial organisms at all levels. Wines are available at major supermarkets, off-licenses and organic retailers. The US based website provides information on the winemaker and his wines and on organic grape farming.
Address: Brown-Forman Wines International, 51 - 56 Mortimer Street, London W1N 8JE
Tel: 020 7323 9332
Fax: 020 7323 5316

Boots Direct
www.wellbeing.com

Specialities: Baby foods; food supplements; vitamins & minerals; bodycare
Website: Organic baby foods and formulas are shown on this extensive health site. Wellbeing is a television channel and a website with on-line shopping. Boots Advantage card may be used.
Credit cards: Most major cards
Tel: 0845 070 8090

Born
www.first-born.co.uk

Specialities: Nappies; babycare; maternity; baby clothes
Website: As well as giving product information, the website covers subjects such as breastfeeding, vaccinations, sleep and childhood ailments. Alternative health, attachment parenting and breast-feeding sections were under development at the time of writing. Most products are made from natural renewable resources, are made by fairtrade workers and have a minimal environmental impact. A wide range of products is catalogued in the on-line shop ranging from nappies to buggies and shoes and clothes.
On-line ordering: Yes
Credit cards: Visa; Mastercard; Switch
Email: info@borndirect.com
Address: 64 Gloucester Road, Bishopston, Bristol BS7 8BH
Tel: 0117 924 5080
Fax: 0117 924 9040

Bowen Association UK
www.bowen-technique.co.uk

Specialities: Bowen technique; organisation; courses
Website: Website provides an introductory explanation of the technique followed by more detailed information on its uses and history. There is a section on courses and training in the UK with venues, dates and fees.
Links: Yes
Contact: John Wilks
Email: office@bowen-technique.co.uk
Address: PO Box 4358, Dorchester, Dorset DT2 7XX
Tel: 0700 269 8324
Fax: 01305 849421

Bowen Technique (Courses)
www.therapy-training.com

Specialities: Bowen technique; courses
Website: Website explains the Bowen technique and provides information on practitioner training. Dates for courses are given and there are details of accredited instructors. There is a printable order form.
Contact: John Wilks
Email: johnwilks@cwcom.net
Address: Therapy Trainings, 742 Corton Denham, Sherborne, Dorset DT9 4LX
Tel: 01963 220615
Fax: 020 8749 6952

Brahma Kumaris World Spiritual University
www.bkwsu.org.uk

Specialities: Spirituality; meditation
Website: A university for understanding the self through meditation and spiritual study. With some 4,000 branches in over 70 countries it offers people of all backgrounds an opportunity to deepen their understanding of universal spiritual principles and study a variety of educational programmes and courses and learning resources. The website gives details of free courses and information on meditation, thinking positively, stress-free living, self-esteem and self-management systems.
Email: london@bkwsu.com
Address: Global Co-operation House, 65 Pound Lane, London NW10 2HH
Tel: 020 8459 1400
Fax: 020 8727 3351

The Brainwave Centre
www.brainwave.org.uk

Specialities: Natural therapies; brain injury
Website: Centre for rehabilitation and development using natural therapies including music for children with brain injuries. Website explains the therapies, gives information on research and case histories. There is a facility for on-line donations to the charity.
Email: brainwave@compuserve.com
Address: Huntworth Gate, Bridgwater, Somerset TA6 6LQ
Tel: 01278 429 089
Fax: 01278 429 622

Brewhurst Health Food Supplies Ltd
www.brewhurst.com

Specialities: Distributor; wholesaler; health food
Website: Major distributor of health food products in the UK. Over 5,500 natural health products from over 250 UK and overseas manufacturers. Website describes commercial retailer scanning device for re-ordering products and gives general background information.
Email: info@brewhurst.com
Address: Abbot Close, Oyster Lane, Byfleet, Surrey KT13 7JP
Tel: 01932 354211
Fax: 01932 353725

Brian Wogan Ltd
www.wogan-coffee.co.uk

Speciality: Coffee
Website: Specialist importer and roaster of organic coffees. Also supplies chemical-free decaffeinated coffees. The on-line shop covers retail coffee packs, blended beans and single origin coffee beans. Website also provides history, coffee machines and coffee product information.
On-line ordering: Yes
Credit cards: Most major credit cards
Email: sales@wogan-coffee.co.uk
Address: Bourbon House, 2 Clement Street, Bristol BS2 9EQ
Tel: 0117 955 3564
Fax: 0117 954 1605

Bristol Cancer Help Centre
www.bristolcancerhelp.org

Speciality: Cancer care
Website: The centre offers healing and positive healthcare to people affected by cancer and to their supporters. It practises, teaches, researches and develops a holistic approach as an integral part of cancer care. The website provides information on the therapy, education, fundraising and nutrition. There is an on-line shop for audio cassettes and CDs, natural bodycare, books, candles, vitamins, supplements and Bristol Cancer Help Centre products.
On-line ordering: Yes
Credit cards: Visa; Mastercard; Access
Email: info@bristolcancerhelp.org
Address: Grove House, Cornwallis Grove, Clifton, Bristol BS8 4PG
Tel: 0117 980 9500
Fax: 0117 923 9184

British Acupuncture Council
www.acupuncture.org.uk

Specialities: Organisation; acupuncture
Website: The Association represents 1,800 acupuncturists. The BAcC maintains standards of education, ethics, discipline and codes of practice. The website provides representative information about all aspects of acupuncture including treatment, practitioners, training, membership, health and safety and research. A list of local practitioners will be supplied free of charge on application.
Email: info@acupuncture.org.uk
Address: 63 Jeddo Road, London W12 9HQ
Tel: 020 8735 0400
Fax: 020 8735 0404

British Allergy Foundation
www.allergyfoundation.com

Specialities: Allergies; organisation
Website: The foundation provides information, advice and support to allergy sufferers and their carers. The site gives information on the foundation, on allergies and allergy testing as well as where to get more help for those who have allergic reactions. There is a page of useful contacts and one on support networks.
Links: Yes
Email: info@allergyfoundation.com
Address: Deepene House,
30 Bellegrove Road, Welling,
Kent DA16 3PY
Tel: 020 08303 8525
Fax: 020 8303 8792

British Association of Homoeopath Veterinary Surgeons
www.bahvs.com

Specialities: Organisation; petcare; animal care; veterinary
Website: The association was formed to advance the understanding, knowledge and practice of veterinary homoeopathy. There are over 150 members. The website explains homoeopathy and gives information on its history and development. Membership details are provided. There is an on-line directory of members listed by county. There is information on events, education and further reading.
Email: enquiries@bahvs.com
Address: c/o Alternative Veterinary Medicine Centre, Stanford-in-the Vale, Oxfordshire SN7 8NQ
Tel: 01367 718115

British Association of Psychotherapists (BAP)
www.bap-pyschotherapy.org

Specialities: Psychotherapy; organisation
Website: The association specialises in individual psychoanalytic psychotherapy and is one of the foremost psychoanalytic psychotherapy training organisations in the UK. The website informs about BAP, its code of ethics, the reduced fee scheme, professional trainings and qualifications, the library, journal and the supervision service.
Links: Yes
Contact: Mrs Elise Ormerod
Email: mail@bap-psychotherapy.org
Address: 37 Mapesbury Road,
London NW2 4HJ
Tel: 020 8452 9823
Fax: 020 8452 5182

British Bakels
www.bakels.com

Specialities: Cake mixes; baking products
Website: Manufacturers of organic cake mixes, scone mixes, muffin mixes, bread improvers, and baking powders. Website provides worldwide details of Bakels.
Contact: P J Hemson
Email: phemsonBB@bakels.com
Address: Granville Way,
off Launton Road, Bicester,
Oxfordshire OX6 0JT
Tel: 01869 247098
Fax: 01869 242979

British Beekeepers Association
www.bbka.org.uk

Specialities: Organisation; beekeeping
Website: The Association represents member associations which in turn represent counties in England. The website provides information on membership, benefits, news, events, the executive committee and BBKA staff. It also lists members associations and addresses in England. There is a section for FAQs about honeybees.
Links: Yes
Catalogue: Information pack (A5 SAE)
Email: information@bbka.demon.co.uk
Address: National Beekeeping Centre, National Agriculture Centre, Stoneleigh Park, Warwickshire CV8 2LG
Fax: 02476 690682

British Chiropractic Association (BCA)
www.chiropractic-uk.co.uk

Specialities: Chiropractic; organisation
Website: The largest and longest established association for chiropractors in the UK. The website is split into sections covering membership information, details for health care professionals, benefits for commerce and industry, chiropractic for everyone, press releases and find-a-chiropractor.
Email: enquiries@chiropractic-uk.co.uk
Address: Blagrave House, 17 Blagrave Street, Reading, Berkshire RG1 1QB
Tel: 0118 950 5950
Fax: 0118 958 8946

British College of Naturopathy and Osteopathy
www.bcno.ac.uk

Specialities: Organisation; naturopathy; osteopathy; natural remedies; courses
Website: Naturopathic or holistic osteopathy focuses on more than diagnosing and treating the structural and mechanical problems of the body through a system of natural therapeutics based on stimulating the body's inherent healing processes. The website provides explanations of the treatments. It also gives details of the courses available and how to apply as well as career information.
Links: Yes
Email: am@bcno.ac.uk
Address: Lief House, 3 Sumpter Close, 120 - 22 Finchley Road, London NW3 5HR
Tel: 020 7435 6464
Fax: 020 7431 3630

British Complementary Medicine Association (BCMA)
www.bcma.co.uk

Specialities: Complementary medicine; organisation
Website: The organisation enables professional standards of complementary medicine in the UK. The website tells the BCMA story and provides information on schools, clinics, therapists and members. There is also an internet health library link. The find-a-therapist facility searches by therapy heading.
Links: Yes
Email: info@bcma.co.uk
Address: PO Box 2074, Seaford, Sussex BN25 1HQ
Tel: 0845 345 5977
Fax: 0845 345 5978

British Homoeopathic Association
www.trusthomeopathy.org

Specialities: Organisation; homoeopathy
Website: The website provides accurate and helpful information about homoeopathy in general. It also gives details of the association and the professional body which regulates the practice of homoeopathy through statutorily registered professionals. It gives information on how to get treatment and its practice in the NHS system. There is a search facility to locate a local practitioner.

Email: info@trusthomeopathy.org
Address: 27a Devonshire Street,
London W1N 1RJ
Tel: 020 7566 7800
Fax: 020 7846 2957

British Holistic Medicine Association
www.bhma.org

Specialities: Holistic medicine; organisation
Website: The aim is to educate doctors, medical students, allied health professionals and members of the general public in the principles and practice of holistic medicine. Website gives information on the association, on holistic medicine and on membership. Self-help tapes are available. There is a diary of events.
Links: Yes
Email: bhma@bhma.org
Address: 59 Lansdowne Place,
Hove, East Sussex BN3 1FL
Tel: 01273 725 951
Fax: 01273 725 951

British Nutrition Foundation
www.nutrition.org.uk

Specialities: Organisation; nutrition
Website: The foundation promotes the nutritional well-being of society through the impartial interpretation and effective dissemination of scientifically based nutritional knowledge and advice. It works with academic and research institutes, the food industry, educators and government. The website provide nutrition facts, news, education, media, events, projects and awards. There is a search facility.
Email: postbox@nutrition.org.uk
Address: High Holborn House,
52 - 54 High Holborn,
London WC1 6RQ
Tel: 020 7404 6504
Fax: 020 7404 6747

British Reflexology Association
www.britreflex.co.uk

Specialities: Reflexology; organisation; courses
Website: Website gives information on the association and on the official teaching body of the association. There are details of the courses with venues, course structure, case-work, examination, duration and applications. There is a listing of publications available for sale as well as charts and rollers and there is a printable order form.
Credit cards: Most major credit cards
Email: bra@britreflex.co.uk
Address: Monks Orchard,
Whitbourne, Worcester,
Worcestershire WR6 5RB
Tel: 01886 821 207
Fax: 01886 822 017

British School of Complementary Therapies
www.bsct.co.uk

Specialities: Complementary therapies; courses; practitioner training
Website: The aim of the school is to provide the highest quality teaching and the most relevant recognition for students wishing to pursue a career or simply follow a course for general interest. All practitioner diplomas are recognised by independent boards. Website gives details of courses, short courses, dates, staff and course recognition. There are courses on reflexology, aromatherapy, massage and anatomy among others. There is an on-line enrolment form with on-line payment facility.
On-line ordering: Yes
Contact: Jane Scrivener
Email: enquiries@bsct.co.uk
Address: 140 Harley Street,
London W1N 1AH
Tel: 020 7224 2394
Fax: 020 7486 2513

British School of Reflexology Sales Ltd
www.footreflexology.com

Specialities: Reflexology; courses; practitioner training
Website: Website gives information on professional reflexology training courses with schedules and course outlines. There is also general information on the school, publications by Ann Gillanders, practitioners and vital products.
Contact: Ann Gillanders
Email: mail@footreflexology.com
Address: Holistic Training Centre, 92 Sheering Road, Harlow, Essex CM17 0JW
Tel: 01279 429 060
Fax: 01279 445 234

British School of Homoeopathy
www.homoeopathy.co.uk

Specialities: Homoeopathy; courses; practitioner training
Website: The mission of the school is to train and graduate highly qualified professional homoeopaths. Website explains about the MSH and gives information on education and practitioner development. There is an enrolment application form which can be printed out. There is contact information for help in finding a homoeopath.
Links: Yes
Contact: Sarah Ainger
Email: learnhomoeopathy@aol.com
Address: Lily Cottage, 5 Townsend, Chittlehampton, Umberleigh, Devon EX37 9PU
Tel: 01769 540155

British School of Osteopathy
www.bso.ac.uk

Specialities: Organisation; osteopathy; courses
Website: Website explains the benefits of osteopathy and the conditions for which it is beneficial. There is information on the school and the. The clinic is described and a fee structure is listed.
Links: Yes
Email: admin@bso.ac.uk
Address: 275 Borough High Street, London SE1 1JE
Tel: 020 7407 0222
Fax: 020 7839 1098

British Society of Clinical Hypnosis
www.bsch.org.uk

Specialities: Organisation; hypnosis
Website: A professional body of hypnotherapists whose aim is to promote and assure high standards in the profession of hypnotherapy. Membership requires high standards of training and ethical practice. From the website it is possible to search for a hypnotist in a given area, learn about the code of conduct expected from members and learn the requirements of membership.
Links: Yes
Email: sec@bsch.org.uk
Address: Queensgate, Bridlington, East Yorkshire YO16 7JQ
Tel: 01262 403103

British Trust for Conservation Volunteers
www.btcv.org.uk

Specialities: Conservation; organisation; holidays
Website: The charity runs 500 conservation holidays a year for volunteers. These include creating community gardens, hedge-laying, coppicing, maintaining footpaths, drystone walling or restoring traditional buildings. The website gives details of volunteering opportunities, community groups and schools, training and jobs, funders and partners.
Email: information@btcv.org.uk
Address: 36 St Mary's Street, Wallingford, Oxfordshire OX10 0EU
Tel: 01491 821600
Fax: 01491 839646

British Wheel of Yoga
www.bwy.org.uk

Specialities: Yoga; organisation
Website: BWY acts as a focus for yoga organisations throughout the UK and provides facilities for people regardless of age, ability, class, colour or creed. Section headings on the website include yoga-for-all, classes, events, training, memberships, insurance, products, magazine, and yoga centre. Full details and dates are given for training courses around the country.
Email: office@bwy.org.uk
Address: 25 Jermyn Street, Sleaford, Lincolnshire NG34 7RU
Tel: 01529 306 851
Fax: 01529 303 233

British Wind Energy Association
www.bwea.com

Specialities: Wind energy; alternative energy; organisation
Website: The largest renewable energy trade association in the UK with a membership of over 500. The website has sections on planning for wind energy, wind farms of the UK, wind power primer, offshore wind energy, education and careers amongst others. There is also an on-line directory of member companies which is searchable both alphabetically and by expertise. There is information on relevant publications.
Links: Yes
Email: info@bwea.com
Address: 26 Spring Street, London W2 1JA
Tel: 020 7402 7102
Fax: 020 7402 7107

Broughton Pastures Organic Wines
www.organicfruitwine.co.uk

Speciality: Fruit wines
Website: The UK's foremost producer of organic fruit wines in a range of flavours: blackcurrant, elderberry, elderflower, ginger, mead and blackberry. Website was under construction at time of writing.
Contact: Brian Reid
Email: organicfruitwine@aol.com
Address: The Old Brewery, 24 High Street, Tring, Hertfordshire HP23 5AH
Tel: 01442 823993

Bumblebee
www.bumblebee.co.uk

Specialities: Wines; cheeses; olive oils; gluten-free products; box scheme; vegetarian
Website: One of London's largest independent retailers of vegetarian, organic and whole foods. Wide range of organic and vegetarian foods with delivery to local postcodes. Box scheme available for North London. Website provides details of the company's background and services.
Contact: Gillian Haslop
Email: bumblebee@virgin.net
Address: 30-33 Brecknock Road, London N7 0DD
Tel: 020 7607 1936
Fax: 020 7607 1936

Buxton Foods
www.stamp-collection.co.uk

Specialities: Flours; chocolate; pasta; wheat-free food; dairy-free foods; sheep cheeses
Website: Producers of the Stamp Collection range of organic, wheat-free, dairy-free food products. The company was founded by the actor Terence Stamp and Elizabeth Buxton. Website provides descriptive listing of products and an email questionnaire to advise customers on their nearest local stockists. There is an FAQ page, a did-you-know page, feedback and trade information.
Links: Yes
Contact: Elizabeth Buxton
Email: e.buxton@stamp-collection.co.uk
Address: 12 Harley Street, London W1N 1AA
Tel: 020 7637 5505
Fax: 020 7436 0979

By Natural Selection
www.bynaturalselection.com

Specialities: Pets; natural remedies for pets; animal care
Website: Stockists of a wide range of alternative treatments for common pet ailments as well as natural healthcare products for pets. There is a detailed, illustrated product listing with on-line ordering facility.
On-line ordering: Yes
Credit cards: Visa; Mastercard; Solo; Switch
Email: sales@bynaturalselection.com
Address: 66 Mountfield Road, Wroxall, Isle of Wight PO38 3BX
Tel: 01983 852 165
Fax: 01983 852 165

Café@Yum
www.yum.org.uk

Specialities: Café; ready meals
Website: A stylish new café bar serving local Welsh and organic snacks for lunch and full meals to take home at night. Website gives details.
Address: 20 West Bute Street, Cardiff Bay CF10 5EP, Wales
Tel: 029 2040 6800

The Caledonian Brewing Co Ltd
www.caledonian-brewery.co.uk

Specialities: Beer; brewery
Website: On the website there is a guided tour of the historic brewery. There is an on-line shopping facility for beers and associated merchandise.
On-line ordering: Yes
Credit cards: Most major credit cards
Contact: Stephen Crawley
Email: info@caledonian-brewery.co.uk
Address: 42 Slateford Road, Edinburgh, Lothian EH11 1PH, Scotland
Tel: 0131 337 1286
Fax: 0131 313 2370

Camphill Village Trust - Botton Village
www.camphill.org.uk

Specialities: Community trust; special needs
Website: A community for adults with special needs based on 5 mixed farms run on bio-dynamic principles. The website is the main site for the Camphill Association which provides places where those with developmental disabilities can live and work with others in an atmosphere of care and respect. Details of the Botton Village Trust are on the website.
Links: Yes
Contact: E Wenneke
Email: botton@camphill.org.uk
Address: Botton Village, Danby, Whitby, North Yorkshire YO21 2NJ
Tel: 01287 660871
Fax: 01287 660888

Canterbury Wholefoods
www.canterbury-wholefoods.co.uk

Specialities: Retail shop; fruit; vegetables; wines; breads; herbal remedies
Website: Free delivery service in Kent for orders over £30. Over 3,000 wholefood lines stocked. Run as a workers co-operative. A general summary of product lines is given on the site together with background information and a map.
Catalogue: Price list available
Contact: Roger Everatt
Email:
info@canterbury-wholefoods.co.uk
Address: Jewry Lane,
Canterbury,
Kent CT1 2NS
Tel: 01227 464623
Fax: 01227 764838

Cariad Aromatherapy
www.cariad.co.uk

Specialities: Aromatherapy products; bodycare; courses
Website: Skincare range for retail and professionals. There is a nationwide beauty training programme. Products are fully and helpfully described on the web pages and there is also an email ordering service. Details are given of the various courses and workshops.
Links: Yes
Contact: Haydn Taylor
Email: info@cariad.co.uk
Address: Cariad Ltd, Rivernook Farm, Sunnyside, Walton on Thames, Surrey KT12 2ET
Tel: 01932 269962
Fax: 01932 253220

Carmichael Estate Farm Meats
www.carmichael.co.uk/venison

Specialities: Meat; venison
Website: Produces traditional meat products from cattle fed on grass and conserved home-grown feed. No hormone implants, growth promoters or other artificial feed additives are used. Website provides on-line ordering. There is also general information on the Carmichael Estates including holidays, visitor centre and sawmill timber products.
Links: Yes
On-line ordering: Yes
Credit cards: Most major credit cards
Email: thechief@carmichael.co.uk
Address: Carmichael, Biggar, Lanarkshire ML12 6PG, Scotland
Tel: 01899 308 336
Fax: 01899 308 481

Casa Paint Company
www.casa.co.uk

Speciality: Paints
Website: Traditionally produced hand-made paints in Mediterranean colours. They contain organic pigments of the highest quality and lots of chalk. There are 22 colours. The website provides information on paint effects and paint outdoors and in the garden. There is a full colour swatch and a colour mixing page. Orders for paint may be made on-line.
Links: Yes
On-line ordering: Yes
Credit cards: Most major cards
Catalogue: Free printed catalogue
Email: enquiries@thebluepenguin.co.uk
Address: PO Box 77, Thame, Oxfordshire OX9 3FZ
Tel: 01296 770139
Fax: 01296 482241

Central London School of Reflexology
www.clsr.clara.net

Specialities: Reflexology; courses; practitioner training
Website: Professional, comprehensive and informative courses with recognised qualifications. Website gives details of the courses and content, and information on the school. A prospectus and information pack is available by return on request.
Catalogue: Prospectus & information pack
Contact: Michael Keet
Email: clsr@clara.net
Address: 15 King Street, Covent Garden, London WC2E 8HN
Tel: 020 7240 1438
Fax: 020 7240 1438

Centre for Alternative Technology
www.cat.org.uk

Specialities: Information; education; sustainable technology; courses; organisation; alternative energy
Website: Promotes practical solutions and information on sustainable technologies. Provides information, educational services and residential courses. Mail order for books and products. The website covers education, special events, virtual tour, information, green shop, news and jobs, consultancy, membership, courses, support, donations and volunteers. Extensive, indexed links page with search facility.
Links: Yes
Contact: Marketing Officer
Email: info@cat.org.uk
Address: Machynlleth,
Powys SY20 9AZ, Wales
Tel: 01654 702400
Fax: 01654 702782

Centre for Pranic Healing
www.pranichealing.co.uk

Specialities: Pranic healing; courses
Website: The website explains the principles of pranic healing which uses 'no touch' methodology. It also provides details and venues of workshops where pranic healing methods may be learnt.
Email: caroline@pranichealing.co.uk
Address: 57 Whitehall Park,
Highgate, London N19 3TW
Tel: 020 7263 4746
Fax: 020 7263 7171

Centres For Change
www.c4c.oxfree.com

Specialities: Organisation; environment; sustainable development
Website: A network of environmental and sustainability centres in the UK. On the website there are newspages and a facility to check out regional centres throughout the UK with addresses and contact details.
Email: cfc@oxfree.com
Address: c/o Friends Meeting House,
43 St Giles, Oxford,
Oxfordshire OX1 3EW
Tel: 01865 316338
Fax: 01865 516288

Chauncey's
www.chauncey.co.uk

Specialities: Building materials; flooring
Website: Reclaimed timber flooring in a range of styles and timbers. The timber gallery on the website allows comparison of styles, colour and grain. There is information on oak, pitch pine and laminate. The pricing page is in £ per square metre. There is information on fitting and details of some recent commercial and domestic projects undertaken. There is a newspage for special offers.
Credit cards: Visa
Email: sales@chauncey.co.uk
Address: 15 - 16 Feeder Road,
Bristol BS2 0SB
Tel: 0117 971 3131
Fax: 0117 971 2224

Cheeky Rascals
www.cheekyrascals.co.uk

Specialities: Baby equipment; nappies
Website: Importers of baby and toddler nursery equipment including washable nappies with national delivery facilities. A collection of practical and innovative products from all over the world. Website was under construction at time of writing.
Credit cards: Visa; Mastercard
Contact: Selina Russell
Email: sales@cheekyrascals.com
Address: The Briars,
Petworth Road, Witley,
Surrey GU8 5QW
Tel: 01428 682488
Fax: 01428 682489

Chiltern Seeds
www.chilternseeds.co.uk

Specialities: Seeds; vegetable seeds
Website: Supplies seeds of all types, including organic vegetable seeds. Website provides extensive catalogue search facility with on-line purchasing. A printed catalogue may be requested on-line. There are a number of pages of information on subjects like greenhouses, pollination, cultivation and endemic.
On-line ordering: Yes
Credit cards: Visa; Mastercard; Switch; Amex
Export: Yes
Catalogue: Yes
Contact: A Bushell
Email: info@chilternseeds.co.uk
Address: Bortree Stile, Ulverston, Cumbria LA12 7PB
Tel: 01229 581137
Fax: 01229 584549

Chops Away
www.chopsaway.com

Specialities: Meat; herbs; plants; holidays; camping
Website: A co-operative of organic and bio-dynamic farmers and growers. The website gives details of family camping on a bio-dynamic farm in South Devon. Many of the campers are Rudolf Steiner School people. Entertainment workshops are organised. A booking form is provided.
Contact: Noni Mackenzie
Email: noni@chopsaway.com
Address: 1 Ticklemore Court, Ticklemore Street, Totnes, Devon TQ9 5EJ
Tel: 01803 864404
Fax: 01803 864404

Clearspring Ltd
www.clearspring.co.uk

Specialities: Non-dairy food; wheat-free foods; Japanese foods; pasta; macrobiotic; vegetarian & vegan
Website: Clearspring's website is divided into three kitchen headings: The Organic Kitchen, The Dietetic Kitchen and the Authentic Japanese Kitchen. They provide own brand and other manufacturers' unprocessed, pure, natural food products available both direct and through retailers. The Japanese Kitchen provides the broadest range of traditionally made Japanese macrobiotic foods available in Europe. All products are made with 100% natural ingredients, GM-free and dairy-free, contain no added sugar and are completely vegetarian/vegan. Extensive website shows products by category, recipes, news, mail order and list of suppliers. There is on-line mail order service direct to the consumer.
Links: Yes
On-line ordering: Yes
Credit cards: Most major cards
Contact: Robert Wilson
Email: mailorder@clearspring.co.uk
Address: 19A Acton Park Estate, London W3 7QE
Tel: 020 8749 1781
Fax: 020 8746 2259

Clearwell Caves
www.clearwellcaves.com

Speciality: Paints
Website: Website contains information on the history of mining for ochre pigments in the Forest of Dean which goes back 4,000 years. Natural ochres to make paint pigments are still produced on a small scale using traditional techniques which have remained unchanged. Courses are run on interior decoration using natural pigments and paints.
Contact: Jonathan Wright
Email: jw@clearwellcaves.com
Address: Nr Coleford, Royal Forest of Dean, Gloucestershire GL16 8JR
Tel: 01594 832 535
Fax: 01594 833 362

Clipper Teas Ltd
www.clipper-teas.com

Specialities: Teas; coffees; fruit teas
Website: Winners of numerous organic food awards, Clipper stretches the boundaries of tea and coffee innovation, providing products without exploitation of people or planet. Teas and coffees may be purchased on-line. There are information pages on organics, fairtrade, teas and coffees.
Links: Yes
On-line ordering: Yes
Credit cards: Visa; Mastercard
Contact: Lorraine Brehme
Email: enquiries@clipper-teas.com
Address: Broadwindsor Road, Beaminster, Dorset DT8 3PR
Tel: 01308 863 344
Fax: 01308 863 847

Club Chef Direct
www.clubchefdirect.co.uk

Specialities: Fish; smoked salmon
Website: Provides recipes and ingredients: fresh seafood, smoked products, wild mushrooms and Asian and fusion ingredients. Some foods are organic. Non-members may place orders but delivery times can only be guaranteed to club members.
Credit cards: Most major cards
Email: info@clubchefdirect.co.uk
Address: Lakeside, Bridwater Road, Barrow Gurney, Bristol BS48 3SJ
Tel: 01275 475252
Fax: 01275 475167

College of Integrated Chinese Medicine
www.cicm.org.uk

Specialities: Acupuncture; Chinese medicine; courses; practitioner training
Website: The CICM trains acupuncture practitioners in an integrated style using the strengths of Yin Yang theory and Five Element constitutional diagnosis in one integrated whole. Website discusses why one should study acupuncture and gives details of undergraduate and postgraduate courses. It also gives a full introduction to the profession, the college, the academic year, assessment, fees, the selection policy and the college administration team.
Email: info@cicm.org.uk
Address: 19 Castle Street, Reading, Berkshire RG1 7SB
Tel: 0118 950 8880
Fax: 0118 950 8890

College of Natural Nutrition
www.natnut.co.uk

Specialities: Nutrition; courses; practitioner training
Website: College courses cover the importance of diet, nutrition and naturopathic techniques in the process of healing. The website provides information on courses, venues, scholarships, lecturers and common questions. There is a newsgroup. Case histories can be submitted for personal consultations for a fee. There is also a facility for finding a practitioner in a given area.
Links: Yes
Email: cnn@globalnet.co.uk
Address: 1 Halthaies, Bradninch, Nr Exeter, Devon EX5 4LQ
Tel: 01392 881 091
Fax: 01392 881 122

College of Natural Therapy
www.colnat.co.uk

Specialities: Natural therapies; naturopathy; nutrition; homoeopathy; courses; practitioner training
Website: The site gives information on natural therapy courses available for home study. It gives the background to the college and its goals and sets out entry requirements, course information, fees, enrolment, terms and conditions, overseas students and application form. The prospectuses for the different subject courses may be printed from the website.
Contact: Norman Eddie
Email: info@colnat.co.uk
Address: 133 Gatley Road, Gatley, Cheadle, Cheshire SK8 4PD
Tel: 0161 491 4313
Fax: 0161 491 4190

College of Naturopathic and Complementary Medicine
www.naturopathy-uk.com

Specialities: Naturopathy; complementary medicine; courses; practitioner training
Website: The website gives a general introduction to naturopathy. There is advice on who can become a naturopath followed by pages on the consultant diploma course, the course syllabus, clinical practice and the diploma. There is an on-line application form and a printable enrolment form.
Contact: Hermann Keppler
Email: info@bestcare-uk.com
Address: 73 Gardenwood Road, East Grinstead, West Sussex RH19 1RX
Tel: 01342 410 505
Fax: 01342 410 909

Coln Valley Smokery
www.colnvalley.co.uk

Specialities: Smoked fish; smoked salmon
Website: Probably the only high quality traditional smokery in commercial production today, selling smoked wild salmon, trout, and eel. Mail order department operates all year round.
On-line ordering: Yes
Credit cards: Most major cards
Email: sales@colnvalley.co.uk
Address: Units 1 & 2, Dovecote Workshops, Barnsley Park, Gloucestershire GL7 5EG
Tel: 01285 740311
Fax: 01285 740411

Colouring Through Nature
www.colouring-thru-nature.co.uk

Speciality: Dyes
Website: The company brings natural dyeing processes into the mainstream with traditional dyes using precise colouring techniques, organic process and a modern twist. It has taken the knowledge of small scale natural dyeing to an industrial level using cutting edge technology. Website was under construction at time of going to press.
Contact: Sushma Patel
Email: info@colouring-thru-nature.co.uk
Address: Unit 16, Old School Works, Narberth, Pembrokeshire SA67 7DU, Wales
Tel: 01834 869042
Fax: 01834 869042

Comfort & Joy
www.comfortandjoy.co.uk

Speciality: Skincare
Website: Healthy, wholesome skincare through products which are gentle, and as unprocessed and natural as possible. Orders can be placed by email, phone or fax. There is information on the site on products for the face, the hair, the body and bathtime as well as for babies. There is a downloadable order form for printing out.
Credit cards: Most major credit cards
Email: merri@comfortandjoy.co.uk
Address: Bay Tree Cottage, East Leach, Nr Cirencester, Gloucestershire GL7 3NL
Tel: 01367 850 278
Fax: 01367 850 278

Community Composting Network
www.othas.org.uk

Specialities: Organisation; composting
Website: A national network promoting the environmental, social and economic benefits of community composting and waste management. Publishes a newsletter, *The Growing Heap*. When opening the website click on the CCN link at the bottom of the page for details of the network.
Links: Yes
Catalogue: Free information pack
Contact: David Middlemass
Email: ccn@gn.apc.org
Address: 67 Alexandra Road, Sheffield, Yorkshire S2 3EE
Tel: 0114 258 0483
Fax: 0114 255 1400

Community Recycling Network
www.crn.org.uk

Specialities: Organisation; recycling
Website: Organisation supporting community based recycling groups engaged in sustainable waste management. There is a summary of the aims of the organisation on the website. There are also pages on member case studies, sharing resources, jobs and vacancies, press releases and regional networks. Membership details are provided.
Links: Yes
Email: info@crn.org.uk
Address: Trelawny House, Surrey Street, Bristol BS2 8BS
Tel: 0117 942 0142
Fax: 0117 942 0164

Compassion in World Farming - CIWF
www.ciwf.co.uk

Specialities: Organisation; farming
Website: Campaigning animal welfare organisation which aims to outlaw cruel factory farming systems. Publishes quarterly magazine, *Agscene*. Website gives a summary of the history and aims of the organisation, details of campaigns, ethical and cruelty-free shopping, education and support.
Links: Yes
Contact: Kerry Cuttin
Email: kerry@ciwf.co.uk
Address: Charles House, 5A Charles Street, Petersfield, Hampshire GU32 3BR
Tel: 01730 264 208

Complementary Medicine Association (CMA)
www.the-cma.org.uk

Specialities: Complementary medicine; organisation
Website: Multi-disciplinary association created for practising complementary medical professionals. Website was under re-construction at the time of writing but there was a find-a-practitioner facility categorised by therapy and region.

Email: info@the-cma.org.uk
Address: The Meridian,
142a Greenwich High Road,
London SE10 8NN
Tel: 020 8305 9571
Fax: 020 8305 9571

CONFOCO UK LTD
www.confocouk.com

Specialities: Processed fruit & vegetable products; food industry supplier
Website: Producer of processed fruit and vegetable products, some organic, for the food industry. Website lists the range of fruit and vegetable products as well as natural extracts.
Contact: Christina Wood
Email: confocouk@confocouk.com
Address: Duncan House, Ripley,
Surrey GU23 6AY
Tel: 01483 211288
Fax: 01483 211388

CONKER SHOE CO
www.conkershoes.com

Speciality: Shoes
Website: Fine, handmade leather shoes from a friendly, family orientated business. Website provides a shoe gallery with style sketches and photographs of the shoes which can be ordered with a variety of soles and colours. It is not yet possible to order shoes on-line except for previous customers whose measurements are kept. Prices are also available in dollars and euros. Existing Conker shoes can be given a complete rebuild.
Email: info@conkershoes.com
Address: 28 High Street, Totnes,
Devon TQ9 5RY
Tel: 01803 862 490
Fax: 01803 866 457

CONSTRUCTION RESOURCES
www.ecoconstruct.com

Specialities: Building materials; paints; solar heating
Website: Britain's first ecological builders' merchant and building centre providing environmentally friendly building and decorating materials. It brings together in central London a unique range of state-of-the-art building products and systems. Website provides sections on the company, what is available, education, services, and product descriptions and uses.
Catalogue: Information leaflets
Email: info@ecoconstruct.com
Address: 16 Great Guildford Street,
London SE1 0HS
Tel: 020 7450 2211
Fax: 020 7450 2212

CONSTRUCTIVE INDIVIDUALS
www.constructiveindividuals.com

Specialities: Architects; self-build
Website: A full architectural service combining sensitive design with ecological awareness. Website gives details of working with clients, building systems and design services. They also run self-build courses.
Email:
design@constructive-l.freeserve.co.uk
Address: Trinity Buoy Wharf,
64 Orchard Place,
London E14 0JW
Tel: 020 7515 9299
Fax: 020 7515 9737

Cooks Delight
www.organiccooksdelight.co.uk

Specialities: Bio-dynamic produce; ethically sourced produce; wines; clothing; baby foods
Website: 100 per cent organic and specialises in ethically sourced, bio-dynamic produce. Mail order service. On-line there is a mission statement, news and events, poems and recipes as well as an on-line shop indexed by product categories.
On-line ordering: Yes
Contact: Rex Tyler
Email: info@organiccooksdelight.co.uk
Address: 360-364 High Street, Berkhamstead, Hertfordshire HP4 1HU
Tel: 01442 863584
Fax: 01442 863702

The Co-Operative Bank
www.cooperativebank.co.uk

Specialities: Bank; financial services
Website: Ethical and iconological banking with full on-line information on all aspects of the banking services. Internet banking facilities. Secure on-line application forms.
Address: PO Box 101, 1 Balloon Street, Manchester M60 4EP
Tel: 08457 212212

Coppercare Products
www.coppercare.co.uk

Specialities: Copper bracelets; homoeopathy
Website: Extensive range of over 300 copper bracelets. There is general information on copper on the web pages together with an illustrated gallery of the bracelets. Contact details are provided.
Contact: Bob Neill
Email: coppercare@btinternet.com
Address: 4 Ryder Court, Saxon Way East, Oakley Hay, Corby, Northamptonshire NN18 9NX
Tel: 01536 747739
Fax: 01536 747057

Cornish Fish Direct
www.cornishfish.com

Speciality: Fish
Website: Fresh fish caught off the Cornish coast using environmentally friendly fishing methods. Fish include wild bass, brill, sardines, Dover sole, John Dory, lemon soles, mackerel, plaice, pollock, turbot, and whiting as well as shellfish. Will advise weekly on seasonability of fish available. Overnight delivery throughout England and Wales. There is a page on environmental considerations.
Links: Yes
Credit cards: Most major cards
Email: dee@cornishfish.com
Address: The Pilchard Works, Tolcarne, Newlyn, Cornwall TR18 5QH
Tel: 01736 332112
Fax: 01736 332442

Cortijo Romero
www.cortijo-romero.co.uk

Specialities: Holidays; courses
Website: Year round alternative holidays in Spain with a wide variety of personal development courses and workshops. Website gives full details of the annual programme and availability of places. It also gives information on course leaders, hiring the Cortijo complex, special offers, the local area and travel. A printed brochure is available and there is also a downloadable PDF version.
Links: Yes
Catalogue: Brochure
Email: bookings@cortijo-romero.co.uk
Address: Little Grove, Grove Lane, Chesham, Buckinghamshire HP5 3QQ
Tel: 01494 782 720
Fax: 01494 776 066

Cosmetics To Go
www.cosmetics-to-go.uk.com

Specialities: Cosmetics; bodycare; perfumes; vegetarian & vegan
Website: Prides itself on products which are both fun and effective, environmentally friendly and tested on human volunteers, not animals. All products are vegetarian and vegan products are labelled.
On-line ordering: Yes
Credit cards: Most major credit cards
Email: admin@cosmetics-to-go.uk.com
Address: PO Box 2150, Hastings, Sussex TN35 5ZX
Tel: 01424 201 202
Fax: 01424 715 793

The Cotswold Gourmet
www.cotswoldgourmet.com

Speciality: Meat
Website: Rare breeds meat producer using natural methods and no growth promoters. Beef. pork. lamb, poultry, bacon and home made sausages. Mail order service. Accredited with rare Breeds Survival Trust. There are price lists on the website together with an email order form.
Credit cards: Most major credit cards
Contact: Judy Hancock & Gary Wallace
Email: goodfood@cotswoldgourmet.com
Address: The Butts Farm, South Cerney, Cirencester, Gloucestershire GL7 5QE
Tel: 01285 862224
Fax: 01285 862224

Cotswold Health Products
www.cotsherb.co.uk

Specialities: Food industry supplier; medicinal herbs; spices
Website: Importers and distributors of medicinal herbs, culinary herbs and spices including herbal teas, herb tinctures and capsules, essential oils and organic herbs and spices. Website provides details of products and prices. There is an organic section. Specialises in bulk sales to the food trade.
Contact: Keidrych Davies
Email: sales@cotsherb.demon.co.uk
Address: 5-8 Tabernacle Road, Wotton-Under-Edge, Gloucestershire GL12 7EF
Tel: 01453 843694
Fax: 01453 521375

Cotswold Speciality Foods
www.cotswoldhoney.demon.co.uk

Specialities: Honey; food industry supplier
Website: Organic honey bottled and processed both for retail and industrial use. Website gives illustrated details of the business.
Contact: L Keys
Email: laurie@cotswoldhoney.demon.co.uk
Address: Avenue 3, Station Lane, Witney, Oxfordshire OX8 6JB
Tel: 01993 703294
Fax: 01993 774227

Cotton Bottoms
www.cottonbottoms.co.uk

Speciality: Nappy service
Website: Home delivery and collection service for laundering modern cotton nappies. Chemical-free, no allergies, no animal testing, eco-friendly. Starter pack and user guide available. Website was under construction at time of writing.
Contact: Joanne Freer
Email: sales@cottonbottoms.co.uk
Address: Unit 12, Broomers Hill Park, Broomers Hill Lane, West Sussex RH20 2RY
Tel: 01798 875300
Fax: 01798 875006

Country Smallholding Magazine
www.countrysmallholding.com

Specialities: Livestock; smallholders; gardening; magazine
Website: Practical monthly magazine for smallholders, poultry keepers and organic gardeners. Website gives subscription and advertising rates. There is also a breeders directory for livestock and a rural organisations directory. There is an information exchange for posting queries to exchange with other smallholders. A beginner's guide to keeping poultry is provided on the site.
Links: Yes
Email: info@countrysmallholding.com
Address: Buriton House, Station Road, Newport, Saffron Walden, Essex CB11 3PL
Tel: 01799 540 922
Fax: 01799 541 367

Country Life
www.country-life.com

Specialities: Food supplements; cosmetics; babycare
Website: American family-owned business. Premium health and beauty collection created from all-natural ingredients without animal testing. Hair, body, beauty and oral care collections. Website gives company profile, new products listing, store locator, and business partners.
Contact: Tom Moses
Email: info@country-life.com
Address: 102-104 Church Road, Teddington, Middlesex TW11 8PY
Tel: 020 8614 1411
Fax: 020 8614 1422

Country Products
www.countryproducts.co.uk

Specialities: Dried fruits; spices; muesli; gluten-free
Website: A family firm selling quality food products, including organic and gluten-free, to the retail, health food and catering trades. Website gives details of product ranges, wholesale and retail.
Contact: Mark Leather
Email: countryproducts@fsbdial.co.uk
Address: 11A Centre Park, Tockwith, York, North Yorkshire YO26 7QF
Tel: 01423 358858
Fax: 01423 358858

The Craniosacral Therapy Association of the UK
www.craniosacral.co.uk

Specialities: Craniosacral therapy; courses; practitioner training; organisation
Website: Website provides an overview of the therapy with an FAQ page and then gives details of courses and professional training available at accredited schools and other institutions. There is a search facility for locating a professional practitioner in a given region and a section on resources with suggestions for reading.
Links: Yes
Email: info@craniosacral.co.uk
Address: 27 Old Gloucester Street, London WC1N 3XX
Tel: 0700 784 735

Cream O'Galloway Dairy Co Ltd
www.creamogalloway.co.uk

Specialities: Ice cream; frozen yogurt
Website: Luxury organic ice cream and frozen yogurt made on the farm in innovative flavours. Website gives details of the organic range and provides links to organic information sites.
Links: Yes
Contact: Wilma Dunbar
Email: info@creamogalloway.co.uk
Address: Rainton, Gatehouse of Fleet, Castle Douglas, Dumfries & Galloway DG7 2DR, Scotland
Tel: 01557 814040
Fax: 01557 814040

Crone's
www.crones.co.uk

Specialities: Cider; juices; vinegar
Website: Award winning organic ciders and apple juices. The website provides history of the company and pages on cider, apple juices and vinegars. It gives details of the range and availability.
Email: info@crones.co.uk
Address: Fairview, Fersfield Road, Kenninghall, Norfolk NR16 2DP
Tel: 01379 687 687
Fax: 01379 688 323

Culpeper Herbalists
www.culpeper.co.uk

Specialities: Herbal remedies; bodycare; cosmetics; aromatherapy products
Website: Only natural ingredients are used in their products and none have been tested on animals. Glass bottles are favoured since they are biodegradable. The on-line shopping facility covers the organic range, aromatherapy, herbs and health, house and home, personal care, bathroom, kitchen, books, videos and gifts. There is a section on herb gardening.
On-line ordering: Yes
Credit cards: Most major credit cards
Email: info@culpeper.co.uk
Address: Head Office, Hadstock Road, Linton, Cambridge, Cambridgeshire CB1 6NJ
Tel: 01223 891 196
Fax: 01223 893 104

Daily Bread Co-operative Ltd
www.dailybread.co.uk

Specialities: Bread; flours; grains; fruit; vegetables; dairy
Website: A supplier of products which offer good value for money and offer a positive benefit to the consumer and the environment. Website provides details of retail shops and the Northampton area delivery service.
Links: Yes
Contact: John Clarke
Email: northampton@dailybread.co.uk
Address: The Old Laundry, Bedford Road, Northampton, Northamptonshire NN4 7AD
Tel: 01604 621531
Fax: 01604 603725

Danmar International
www.takeitfromhere.co.uk

Specialities: Italian products; olive oils
Website: This supplier of the very highest quality Italian foods and wines to restaurants and food houses is now able to supply direct to the public through the website. The site has section headings for the cellar, the pantry gift boxes, special offers and recipes. The range includes a number of organic products as well as pasta, pasta sauces, cheese, oils, vinegars, truffle products and wines. (see also Take It From Here)
Links: Yes
On-line ordering: Yes
Credit cards: Visa; Mastercard; Switch; Solo; Amex
Email: sales@tifh.co.uk
Address: Unit 04, BetaWay, Thorpe Industrial Estate, Egham, Surrey TW20 8RX
Tel: 01784 477 812
Fax: 01784 477 813

Dartington Tech - The Regional Centre for Organic Horticulture
www.rcoh.co.uk

Specialities: Courses; resource centre; gardening
Website: The centre offers courses for both professional and amateurs in organic gardening and horticulture. It provides an education and resource centre. Website gives details of the courses available and the qualifications achievable. There is also a page detailing objectives.
Email: info@dartingtontech.co.uk
Address: Westminster House, 38/40 Palace Avenue, Paignton, Devon TQ3 3HB
Tel: 01803 867693
Fax: 01803 867693

David Clarke Associates
www.dcalondon.com

Speciality: Architects
Website: Architects, designers and energy consultants specialising in low energy and solar design. Website provides details of various projects in this field.
Email: dca@compuserve.com
Address: Toll House Studio, Cambridge Cottages, Kew Green, Surrey TW9 3AY
Tel: 020 8332 9696
Fax: 020 8332 2626

Dawson & Son - Wooden Toys and Games
www.dawson-and-son.com

Speciality: Wooden toys
Website: Over 300 wooden toys and games available. These are categorised on the site under active play, early years, educational, furniture, games, older, vehicles and stocking fillers. All are illustrated and described and they may be ordered on-line.
Links: Yes
On-line ordering: Yes
Credit cards: Most major credit cards
Catalogue: Yes
Contact: Nigel Dawson
Email: toys@dawson-and-son.com
Address: Winstanley House, Market Hill, Saffron Walden, Essex CB10 1HQ
Tel: 01799 526 611
Fax: 01799 525 522

The Day Chocolate Company
www.divinechocolate.com

Speciality: Chocolate
Website: Fairly traded milk chocolate. The website explains the history of the company and the principles of fairtrade. There are pages on Ghana and cocoa, how the chocolate is made, and how to get involved. There is also information from an educational point of view. The chocolate is available in fairtrade shops, supermarkets and independent retailers.
Links: Yes
Email: info@divinechocolate.com
Address: 4 Gainsford Street, London SE1 2NE
Tel: 020 7378 6550
Fax: 020 7378 1550

Dead Sea Spa Magik
www.findershealth.com

Specialities: Bodycare; skincare
Website: A natural skin care company specialising in mineral therapy developed from the Dead Sea in Israel. It includes Hydrofloat Spa Therapy using pure Dead Sea Waters - the ultimate relaxation experience. There is a full description with technical information on the web pages.
Contact: Michelle Hodge
Email: spahouse@findershealth.com
Address: Finders International Ltd, Orchard House, Winchet Hill, Kent BR3 6NR
Tel: 01580 211055
Fax: 01580 216062

Defenders Ltd
www.defenders.co.uk

Specialities: Biological controls; gardening; pest controls
Website: A wide selection of tried and tested biological pest controls compatible with the needs of organic gardeners. Website provides comprehensive advice and information on individual pests and how to combat them biologically. Prices are provided. There is a page of helpful tips and advice. A price list and order form can be downloaded.
Email: help@defenders.co.uk
Address: Occupation Road, Wye, Ashford, Kent TN25 5EN
Tel: 01233 813633
Fax: 01233 813633

Delfland Nurseries
www.delfland.co.uk

Specialities: Gardening; wholesaler
Website: Brassicas, celery, leeks, lettuces, Chinese lettuces, salads and herbs in plastic cell trays or semi-discrete blocks. The site was under construction at time of going to press.
Contact: Jill Vaughan
Email: delfland@ndirect.co.uk
Address: Benwick Road, Doddington, March, Cambridgeshire PE15 0TU
Tel: 01354 740553
Fax: 01354 741200

Denes Natural Pet Care
www.denes.com

Specialities: Pet food; petcare; animal care
Website: Producers of natural cat and dog foods and petcare products including herbal medicines, aromatherapy products and supplements that can help improve pets' general health and overall fitness. Details of the different food and other products are given on the website with full information on suitability, ingredients, features and benefits. There is a printable order form for the home delivery service. Free pet care advice on health and feeding patterns can be given.
Links: Yes
Credit cards: Most major credit cards
Email: info@denes.com
Address: 2 Osmond Road, Hove, East Sussex BN3 1BR
Tel: 01273 325704
Fax: 01273 325364

The Deodorant Stone
www.deodorant-stone.co.uk

Specialities: Deodorant stones; bodycare
Website: Crystal deodorant stones are made from natural mineral salts. Stones are hypo-allergenic, non-toxic, unscented, cruelty-free and non-staining. Mail order prices are given on the website. There is also natural health news with wide range of topical articles. Information on wholesale enquiries is provided.
Credit cards: Most major credit cards
Email: info@deodorant-stone.co.uk
Address: 2 Lime Tree Cottage, Foxley, Malmesbury, Wiltshire SN16 0JJ
Tel: 01666 826515
Fax: 01666 823186

Deverill Trout Farm & Purely Organic Farm Shop
www.purelyorganic.co.uk

Specialities: Trout; farm shop; watercress
Website: Organic trout reared in spring water flowing from organic watercress beds. Also fresh watercress, vegetables, fruit, groceries, dairy products and meat available from farm shop. At time of writing orders for trout may phoned through.
Credit cards: Visa; Switch; Mastercard
Email: order@purelyorganic.co.uk
Address: Longbridge Deverill, Warminster, Wiltshire BA12 7DZ
Tel: 01985 841093
Fax: 01985 841268

Discovery Initiatives
www.discoveryinitiatives.com

Specialities: Holidays; activity holidays
Website: Nature travel in association with leading conservation organisations. The website provides many different ways of searching for an appropriate holiday: by country, by animal, by habitat and by activity. There is also a search facility by type of trip, by comfort zone and by activity zone. There is information on charitable challenges. An availabilities page gives the latest news on available places on individual holidays.
Links: Yes
Contact: Julian Matthews
Email: enquiry@discoveryinitiatives.com
Address: The Travel House, 51 Castle Street, Cirencester, Gloucestershire GL7 1QD
Tel: 01285 643333

Distinctive Drinks Company
www.distinctivedrinks.com

Speciality: Beers
Website: A range of organic beers from the Pinkus Muller Brewery in Germany, the world's first certified organic brewery. Also organic lager brewed in the UK. The organic beer range is illustrated and described on the web pages and there is a family tree of the Muller family.
Contact: Ian Haffety
Email: ddc@distinctivedrinks.com
Address: Old Dairy, Broadfield Road, Sheffield, South Yorkshire S8 0XQ
Tel: 0114 255 2002
Fax: 0114 255 2005

Dittisham Fruit Farm
www.self-cater.co.uk/dff

Specialities: Meat; herbs; self-catering accommodation; holidays
Website: Self-catering accommodation available on small, pork producing organic farm. Website gives details of accommodation, photographs, tariffs, booking details and booking form. There are also details of produce available with prices.
Contact: Sue Fildes
Email: suefildes@supanet.com
Address: Capton, Dartmouth, Devon TQ6 0JE
Tel: 01803 712452
Fax: 01803 712452

Divine Wines
www.divinewines.co.uk

Specialities: Wines; beers; balsamic vinegar; olive oils; cordials
Website: Policy of selling excellent organic wines at affordable prices. There is an on-line search facility by taste, colour, region and price. Supplies to the trade and also to individual customers by mail order. Over 120 100% fully organic wines, beers, spirits and fruit cordials are available. On-line ordering facility.
On-line ordering: Yes
Credit cards: Most major cards, debit preferred
Catalogue: Brochure
Contact: Julie Pearson
Email: enquiries@divinewines.co.uk
Address: Divine House, 6 Hornbeam, Leighton Buzzard, Bedfordshire LU7 8UX
Tel: 01525 218100
Fax: 01525 218100

DLW Residential Floorings
www.dlw.co.uk

Specialities: Flooring; linoleum
Website: Manufacturer of various floorings including linoleum. The ingenious website room-viewer offers the opportunity of seeing a choice of floor in a variety of settings.
Address: Centurion Court, Milton Park, Abingdon, Oxfordshire 0X14 4RY
Tel: 01235 831296
Fax: 01235 444001

The Domestic Fowl Trust
www.mywebpage.net/domestic-fowl-trust

Specialities: Poultry housing; poultry; ducks; geese
Website: Main aim is the conservation of rare breeds of domestic poultry, chickens, ducks and geese. Also supplies poultry housing and equipment which are illustrated on the website and in the printed catalogue.
Catalogue: Free mail order catalogue
Contact: Bernie & Clive Landshoff
Email:
domestic-fowl-trust@mywebpage.net
Address: Honeybourne Pastures, Honeybourne, Evesham, Worcestershire WR11 5QG
Tel: 01386 833083
Fax: 01386 833364

Double Dragon
www.doubledragon.co.uk

Specialities: Herbal remedies; food supplements; teas
Website: High quality ginseng roots, slices, extracts, vegicaps and solutions. Also organic green teabags, jasmine flavoured green tea and pure shredded ginseng in teabags. Products are described and illustrated on the website. There is a price list.
Contact: Alice Chiu
Email: info@doubledragon.co.uk
Address: Man Shuen Hong Kong Ltd, 4 Tring Close, Barkingside, Essex IG2 7LQ
Tel: 020 8554 3838
Fax: 020 8554 3883

Doves Farm Foods
www.dovesfarm.co.uk

Specialities: Cereals; flours; bread
Website: Suppliers of breads, home baking flours, cookies and breakfast cereals using the finest organic ingredients for the sake of nutritional quality, sustainability and the environment. Also products in both organic and wheat-free/gluten-free categories. This is a simply constructed website.
Contact: Clare Marriage
Email: mail@dovesfarm.co.uk
Address: Salisbury Road, Hungerford, Berkshire RG17 0RF
Tel: 01488 684880
Fax: 01488 688235

Dr Edward Bach Centre
www.bachcentre.com

Speciality: Herbal remedies
Website: Centre for Bach Flower remedies. Education and information on Bach's work, including publications, special courses for practitioners and referral to practitioners. Details of the centre, the foundation and the trust are all given on the website. The site also gives a list of the remedies and their indications as well as case studies. Information is provided on how to obtain the remedies by mail order.
Links: Yes
Email: mail@bachcentre.com
Address: Mount Vernan, Baker's Lane, Sotwell, Oxfordshire OX10 OPZ
Tel: 01491 834678
Fax: 01491 825022

Dr Hauschka Skin Care
www.drhauschka.co.uk

Speciality: Skincare
Website: Distributors of the Dr Hauschka skin care range of holistic products made from organically grown herbs and plants from certified bio-dynamic farms, A natural and holistic range it treats specific skin conditions rather than skin types. A mail order catalogue is available through the email request form on the site. Full on-line ordering is also available from indexed illustrated order forms. Details of ingredients are given. List of stockists and aestheticians is also provided.
On-line ordering: Yes
Catalogue: Yes
Contact: Yvonne Rowse
Email: enquiries@drhauschka.co.uk
Address:
19-20 Stockwood Business Park,
Stockwood, Nr Redditch,
Worcestershire B96 6SX
Tel: 01386 792622
Fax: 01386792623

The Duke of Cambridge
www.singhboulton.co.uk

Speciality: Pub
Website: Organic gastro-pub serving regional European style organic food and organic beers. Also serves organic baby food. Committed to ethical work practices and organic policies. Sample menus with prices are given on the website. There is an information page on organic drinking. In the section on organics there are pages on the nature of organics, certification, organic policy, organisations and organic pubs.
Contact: Esther Boulton & Geetie Singh
Email: eb@singhboulton.co.uk
Address: 30 St Peter's Street,
London N1 8JT
Tel: 020 7359 1877
Fax: 020 7359 5877

Earthbound
www.earthbound.co.uk

Specialities: Skincare; soaps
Website: Toiletries prepared from organic ingredients and oils. Everything is made by hand. Mail order facility is available and the order form may be printed out from website. Website describes the creams and oils and also provides a photo gallery.
Export: EEC
Contact: Jo Ordonez-Sampson
Email: sales@earthbound.co.uk
Address: The Toll House,
Dolua, Llandrindod Wells,
Powys LD1 5TL, Wales
Tel: 01597 851157
Fax: 01597 851157

The Earth Centre
www.earthcentre.org.uk

Specialities: Eco-park; theme park
Website: Ecological park providing education, leisure, entertainment and enterprise for a sustainable future in 400 acres of environmental theme park for families with events and activities. Website provides opening hours and admission prices, location instructions and a summary of what the Earth Centre is about.
Address: Denaby Main, Doncaster,
South Yorkshire DN12 4EA
Tel: 01709 512933
Fax: 01709 512010

Earth Friendly Products
www.greenbrands.co.uk

Specialities: Household products; cleaning products; vegan
Website: A complete range of natural household laundry and cleaning products which are non-toxic and safe for the environment. Concentrated formulas are biodegradable, vegan and have not been tested on animals An e-brochure can be downloaded from the site. There is secure on-line ordering. Testimonials are provided.

On-line ordering: Yes
Contact: Lois Clark
Email: info@greenbrands.co.uk
Address: Park Lodge,
96 Court Road, Tunbridge Wells,
Kent TN4 8EB
Tel: 01892 616871
Fax: 01892 616238

Earthrise
www.earthrise.com

Specialities: Wheatgrass products; food supplements
Website: Importer of *Earthrise*, a range of spirulina, chlorella, barley and wheatgrass tablets and powders. Website gives information on the Earthrise company, products, spirulina cultivation and safety.
Contact: Neil Dickson
Email: info@all-seasons.demon.co.uk
Address: 19-21 Victoria Road North,
Southsea, Hampshire PO5 1PL
Tel: 02392 755660
Fax: 02392 755660

Earthwatch
www.earthwatch.org/europe

Specialities: Environment; organisation
Website: The mission of Earthwatch is to promote sustainable conservation of our natural resources and cultural heritage by creating partnerships between scientists, educators and the general public. This is accomplished through research, education and conservation. The extensive website provides a considerable range of information on their activities, including details of expeditions.
Email: info@earthwatch.org.uk
Address: 57 Woodstock Road,
Oxford, Oxfordshire OX2 6HJ
Tel: 01865 318 838
Fax: 01865 311 383

Earthwise Baby
www.earthwisebaby.com

Specialities: Parenting; nappies; baby products;
Website: Extensive site. Orders can be placed by secure server, phone, fax or post. A wide range of baby, parenting and mothering products are available. There are about-parenting,about-pregnancy and avout-birth sections. There are also free birth announcements, classified, reseller programme and parenting news. A separate section of the site is a Mind Body Spirit store.
On-line ordering: Yes
Export: Yes
Catalogue: Yes
Contact: S Erlick
Email: sales@earthwisebaby.com
Address: Apsley Distribution Ltd,
PO Box 1708, Apsley Guise,
Milton Keynes,
Buckinghamshire MK17 8YA
Tel: 01908 588771
Fax: 01908 585771

Eastbrook Farm Organic Meats
www.helenbrowningorganics.co.uk

Speciality: Meat
Website: Organic meat, bacon, cured ham and sausages supplied nationwide to retailers and supermarkets and direct to the public through home delivery service. Website provides information on the farm and its aims and why organic eating is desirable. Details are given of products available in supermarkets and there is also an on-line price list and order form for home delivery service.
Contact: Helen Browning
Email:
info@helenbrowningorganics.co.uk
Address: Eastbrook Farm,
Bishopstone, Nr Swindon,
Wiltshire SN6 8PW
Tel: 01793 790460
Fax: 01793 791239

Eco Babes
www.eco-babes.co.uk

Specialities: Nappies; baby clothes; babycare; household products
Website: Eco-babes is committed to finding the right nappy systems or system for the individual. The website lists the full range on its on-line catalogue which also includes organic baby clothes, bedding, hemp slings and organic baby toiletries.
On-line ordering: Yes
Credit cards: Visa; Mastercard; Switch
Contact: Dyane Cakebread
Email: cakebread@eco-babes.co.uk
Address: 79 Orton Drive, Witchford, Ely, Cambridgeshire CB6 2JG
Tel: 01353 664 941

Eco Clothworks
www.clothworks.co.uk

Specialities: Textiles; clothing; bedding; nappies; skincare
Website: Clothes for adults and children made from organic wool, cotton and hemp. There is a full catalogue on-line. The aim of the company is to offer information and safe products. Until e-commerce is fully in place orders can be taken over the phone or by post. There are information pages on organic cotton and hemp.
On-line ordering: Under development
Credit cards: Most major cards
Contact: Linda Row
Email: info@clothworks.co.uk
Address: PO Box 16109, London SE23 4WA
Tel: 020 8299 1619
Fax: 020 8299 6997

The Eco House
www.ecohouse.org.uk

Specialities: Ecological housing; housing
Website: The eco-house is an environmental showhome, using the latest energy saving devices. Its wide range of environment ideas and exhibitions are designed to encourage visitors to try green living for themselves. Admission is free.
Address: Environ, Parkfield, Western Park, Leicester LE3 6AH
Tel: 0116 254 5489
Fax: 0116 255 2343

Eco Solutions Ltd
www.ecosolutions.co.uk

Speciality: Paint removal
Website: Producers of safe and environmentally friendly solvent-free coating removal products without the risks associated with hazardous solvents. Website provides information on the product range and uses together with a guide to retailers, web-based retailers and mail order outlets. There is information on health, safety and the environment.
Links: Yes
Email: homestrip@ecosolutions.co.uk
Address: Summerleaze House, Church Road, Winscombe, Somerset BS25 1BH
Tel: 01934 844 484
Fax: 01934 844 119

Ecobaby
www.ecobaby.ie

Specialities: Nappies; babycare
Website: Leading supplier of eco-friendly baby products. Website provides information on reusable and disposable nappies and gives details of environmental factors and disposal. Experiments are being held in composting nappies.
Email: info@ecobaby.ie
Address: 8 Maltings Business Park, Marrowbone Lane, Dublin 8, Ireland
Tel: + 353 1 415 0877
Fax: + 353 1 415 0878

Ecobrands Ltd
www.ecobrands.co.uk

Specialities: Herbal remedies; aromatherapy products
Website: Company specialises in marketing a range of products to enable users to make positive changes to health and well-being. These include insect bite remedies, aromist therapeutics, travel products, medical products, sports essentials, professional foot care and oral health. Web pages give detailed product information and provide an on-line ordering facility.
On-line ordering: Yes
Contact: Hanan Kandili
Email: hanan@ecobrands.co.uk
Address: 3 Adam & Eve Mews, Kensington, London W8 6UG
Tel: 020 7460 8101
Fax: 020 7565 8779

The Ecological Design Association
www.edaweb.org

Specialities: Design; ecology
Website: The Association links designers of all disciplines in a global network providing information, education, exchange of ideas and stimulation to encourage and sustain the practices of ecological design. Website gives information on the aims and philosophy, membership, services and publications.
Email: ecological@designassociation.freeserve.co.uk
Address: c/o The Ecological Design Association, The British School, Slad Road, Stroud, Gloucestershire GL5 1QW
Tel: 01453 765 575
Fax: 01453 754 211

The Ecologist
www.theecologist.com

Specialities: Magazine; ecology
Website: The world's longest running environmental magazine, founded in 1970, covers broad subject areas with radical analysis of topical and general issues. Website gives details of the current and back issues, campaigns, subscriptions and advertising. Current and back issues may be ordered on-line.
Links: Yes
On-line ordering: Yes
Email: theecologist@galleon.co.uk
Address: Galleon Ltd, PO Box 326, Sittingbourne, Kent ME9 0SA
Tel: 01795 414 963
Fax: 01795 414 555

Ecology Building Society
www.ecology.co.uk

Specialities: Building society; housing
Website: Dedicated to improving the environment by promoting sustainable housing and communities. Savings placed with the society are used to finance organic, healthy food, building energy and resource efficient homes and rescuing derelict houses.
Email: info@ecology.co.uk
Address: 18 Station Road, Cross Hill, Keighly, West Yorkshire BD20 7EH
Tel: 0845 674 5566

Ecomerchant
www.ecomerchant.demon.co.uk

Specialities: Building materials; paints
Website: Green building products, including paints, locally sourced bricks and tiles, insulation and reclaimed or new flooring. On the website there is an on-line product catalogue with ordering facility.
Links: Yes
Address: The Old Filling Station, Head Hill Road, Goodnestone, Nr Faversham, Kent ME13 9BY
Tel: 01795 530130
Fax: 01795 530430

ECONOPACK
www.gardencomposters.com

Specialities: Composting; gardening
Website: A garden composter with twin walls to speed up composting which provides easy access from all sides. Website shows cutaway design of the composter and gives specifications. It also provides assembly instructions and an email form for requesting further information or ordering.
Catalogue: Free composting tips
Email: info@econopack.co.uk
Address: Heage Road Industrial Estate, Ripley, Derbyshire DE5 3FU
Tel: 01773 746285

ECOPINE
www.ecopine.co.uk

Speciality: Furniture
Website: Suppliers of high quality eco-friendly old pine furniture and tables. Everything is made from reclaimed old pine timber. Website provides information on the company, the woods and the designs and illustrates some of the furniture items. There is an email facility for requesting sales literature.
Links: Yes
Catalogue: Brochures
Email: info@ecopine.co.uk
Address: Freepost SEA10514, Wadhurst, Tunbridge Wells, Kent TN5 6BR
Tel: 01892 785 072
Fax: 01892 785 072

ECO-SCHOOLS PROGRAMME
www.eco-schools.org.uk

Specialities: Education; organisation
Website: A programme for schools to get the whole school community working together to improve the environment, encourage citizenship and promote healthy lifestyles. Website gives background information on the programme and the process, has a resources page and explains how to join in.
Email: eco-schools@tidybritain.org.uk
Address: Tidy Britain Group. Elizabeth House The Pier, Wigan, Lancashire WN3 4EX
Tel: 01942 824 620
Fax: 01942 824 778

ECOTRICITY
www.ecotricity.co.uk

Speciality: Alternative energy
Website: Ecotricity founded the UK's green electricity market in 1996 and is now the market leader in Europe. Site was under construction at time of writing.
Contact: Vicki Evans / Clare Summers
Email: info@ecotricity.co.uk
Address: Axiom House, Station Road, Stroud, Gloucestershire GL5 3AP
Tel: 01453 756111
Fax: 01453 756222

ECOVER
www.ecover.com

Specialities: Household products; cleaning products
Website: Ecover produces an ecologically sound range of effective cleaning products. The ingredients come from natural sources and provide the least possible burden to the environment. Products are manufactured at the world's only ecological factory in Belgium which recycles its own waste and runs on minimum energy. The plant has a grass covered roof which provides cooling in the summer and warmth in the winter. The company and its products are described on the website. There is a virtual guided tour of the ecological factory.
Links: Yes
Contact: Clare Allman
Email: soapmaster@ecover.com
Address: 165 Main Street, Newbury, Berkshire RG19 6HN
Tel: 01635 528240
Fax: 01635 52827

Eco-Zone Ltd
www.sea-vegetables.co.uk

Specialities: Seaweed based products; food supplements; bodycare
Website: Fresh and dry seaweed and a range of seaweed based products. Dry or fresh seaweed can be used as everyday cooking ingredients even in western style cooking. Natural food supplements (based on seaweed or micro-algae) and a full range of high quality marine cosmetics are also available. The innovative seaweed based products are all described on the web pages
Contact: Eric Vagniez
Email: ecozone@sea-vegetables.co.uk
Address: 12 Snarsgate Street, London W10 6QP
Tel: 020 8962 6399
Fax: 020 8962 6399

ECOZONE.CO.UK
www.ecozone.co.uk

Specialities: Household; health products; bodycare; wine; babycare; gardening
Website: On-line shopping for eco-products, particularly chemical-free cleaners and cloths, hard water solutions, eco-gadgets, hair, body and beauty, health, organic wine and beer, electrical, pets and garden products.
On-line ordering: Yes
Credit cards: Visa; Mastercard; Switch; Delta
Catalogue: Yes
Contact: Gina Van der Molen
Email: info@ecozone.co.uk
Address: Birchwood House, Briar Lane, Croydon, Surrey CRO 5AD
Tel: 020 8777 3121
Fax: 020 8777 3393

Eden Project
www.edenproject.com

Specialities: Eco park; environment
Website: The Eden Project is working with others toward the implementation and understanding of the principles of Agenda 21, the framework for sustainable development. There is information on partnerships, the Eden Institute, Eden friends, education, finance and sustainable development. Downloads include a shop order form and media information packs on the Eden Project. There is a children's page.
Address: Bodelva, St Austell, Cornwall PL24 2SG
Tel: 01726 811911
Fax: 01726 811912

Education Otherwise
www.education-otherwise.org

Specialities: Home education; organisation
Website: UK based membership organisation providing support and information for families whose children are being educated outside school. Website gives information on the organisation, legal information, educational resources, about home education and publications listings. Postal enquiries should be accompanied by an SAE. There is a printable form for joining as well as an on-line facility with a questionnaire.
On-line ordering: Yes
Credit cards: Visa; Mastercard
Address: PO Box 7420, London N9 9SG
Tel: 0900 1518 303

ELECTRIC CAR ASSOCIATION
www.eca-uk.freeserve.co.uk

Specialities: Electric cars; organisation; alternative energy
Website: Association founded to offer information, advice and support to electric car buyers, users, builders and converters. It is possible to recharge the batteries of electric cars with solar, hydro or wind power generators. Website has pages on what's available and members' cars as well as technical information.
Links: Yes
Email: webmaster@electric-cars.org.uk
Address: Blue Lias House, Station Road, Hatch Beauchamp, Somerset TA3 6SQ
Tel: 01823 480 196
Fax: 01823 481 116

ELLIE NAPPY COMPANY
www.elliepants.co.uk

Speciality: Nappies
Website: Easy to use waterproof covers for washable nappies. The company also sells other baby products including organic baby pyjamas. There are pages of FAQs, tips and testimonials.
On-line ordering: Yes
Credit cards: Most major credit cards
Contact: Lizzie Jewkes
Email: info@elliepants.co.uk
Address: PO Box 16, Ellesmere Port, Cheshire CH66 2BE
Tel: 0151 200 5012
Fax: 0151 339 1464

ELLIS ORGANICS
www.eatorganic.co.uk

Specialities: Meat; poultry; vegetables; fruit; bread; box scheme
Website: Suppliers of home-grown fruit and vegetables and organic meats. Product range includes salads, breads, pasta, dairy and tofu products, bottled and canned foods and confectionery items. Delivery to parts of Berkshire and south Oxfordshire only. Ordering by phone, fax or email with interactive weekly order form.
Links: Yes
Contact: Aidan Carlisle
Email: ellis-organics@clara.net
Address: 5 Lea Road, Sonning Common, Reading, Berkshire RG4 9LH
Tel: 0118 972 2826
Fax: 0118 972 2826

ELM FARM RESEARCH CENTRE
www.efrc.com

Specialities: Farming; research; agriculture
Website: The aim is the development and promotion of organic agriculture as the most environmentally sound way of producing healthy food. EFRC is based on a working farm which provides a basis for research and educational work. The website gives information on research and education as well as news, comment and policy publications.
Links: Yes
Email: elmfarm@efrc.com
Address: Hamstead Marshall, Newbury, Berkshire RG20 0HR
Tel: 01488 658 298
Fax: 01488 658 503

EMERSON COLLEGE
www.emerson.org.uk

Specialities: Courses; bio-dynamics; gardening; agriculture
Website: International centre for adult education based on the work of Rudolf Steiner. One week introductory summer course in bio-dynamic gardening and agriculture. Also 4 year full time training in bio-dynamic agriculture. Website provides dates and other details of short, summer and full time courses available.
Links: Yes
Contact: Belinda Hammond
Email: mail @emerson.org.uk
Address: Hartfield Road, Forest Row, East Sussex RH18 5JX
Tel: 01342 822238
Fax: 01342 826055

The Empty Homes Agency
www.emptyhomes.com

Speciality: Housing
Website: For every homeless person in England there are 7 empty homes. This is an independent charity which highlights the disgrace of empty, wasted and underused homes and property throughout England. The website provides statistics, the national picture, community action, events and reports. There is a regional analysis of the problem.
Email: info@emptyhomes.com
Address: 195 - 197 Victoria Street, London SW1E 5NE
Tel: 020 7828 6288
Fax: 020 7828 7006

Energy Conservation & Solar Centre (ECSC)
www.ecsc.org.uk

Specialities: Energy; organisation
Website: The home of energy advice for tenants, local authorities and housing associations, charities and companies. Dedicated to the development and implementation of programmes to create socially responsible energy use. The website was under reconstruction at the time of writing.
Email: john.thorp@ecsc.org.uk
Address: Unit 327, 30 Great Guildford Street, London SE1 0HS
Tel: 020 7922 1660
Fax: 020 7771 2344

Energy Development Co-operative Ltd
www.unlimited-power.co.uk

Specialities: Solar power; wind power; alternative energy
Website: The core business of this workers' co-operative is supplying renewable energy technologies. It designs prototype solar products or stand-alone energy systems using solar and wind energy as the power source for use anywhere around the world. The website gives background information and has sections on the catalogue, services, questions, downloads and ordering.
Address: Bridge Road, Oulton Broad, Suffolk NR32 3LR
Tel: 01502 589407
Fax: 01502 589120

Energy Saving Trust
www.est.org.uk

Specialities: Organisation; energy; alternative energy
Website: Identifies, publicises and manages energy efficiency programmes. Promotes and advises on all aspects of energy efficiency, renewable energy and clean fuels. The website provides information on a variety of products which save energy inside and outside the home. It also gives details of grants available. There is a savings calculator.
Email: info@est.co.uk
Address: 21 Dartmouth Street, London SW1H 9BP
Tel: 020 7222 0101
Fax: 020 7654 2444

English Hurdle
www.hurdle.co.uk

Specialities: Willow products; gardening
Website: Many gardening products made from willow including hurdles for fencing, plants supports, trellises, arbours, seats and plant climbers made from both living and dead woods. Website gives information on care and maintenance.
On-line ordering: Yes
Credit cards: Most major credit cards
Contact: James Hector
Email: sales@hurdle.co.uk
Address: Curload, Stoke St Gregory, Taunton, Somerset TA3 6JD
Tel: 01823 698 418
Fax: 01823 698 859

English Nature
www.english-nature.co.uk

Specialities: Conservation; organisation; wildlife
Website: Organisation championing the conservation of wildlife and natural features throughout England. Website provides information on the agency and its work, on science and research, publications and maps and on special sites. There is a region by region facility for locating nature reserves.
Links: Yes
Email: enquiries@english-nature.org.uk
Address: Northminster House, Peterborough, Cambridgeshire PE1 1UA
Tel: 01733 455 000
Fax: 01733 568 834

Environ
www.environ.org.uk

Specialities: Environment; organisation
Website: Independent charity working to improve the environment and the communities we live in. It provides information, advice and practical help to encourage individuals and organisations to take practical steps towards a more sustainable future. The website is used as a means of sharing information and ideas. It covers issues, news, events and what the individual can do.
Links: Yes
Email: info@environ.org.uk
Address: Parkfield, Western Park, Leicester, Leicestershire LE3 6HX
Tel: 0116 222 0255
Fax: 0116 255 2343

Environmental Construction Products
www.ecoproducts.co.uk

Specialities: Building materials; paints
Website: Green building materials including their own range of environmentally friendly and energy conserving windows, doors and conservatories. Website has sections on windows, doors, conservatories, glazing, paints, wood finishes, preservatives and other products. There is a newsletter facility.
Links: Yes
Contact: Brochure
Email: info@ecoproducts.co.uk
Address: 26 Millmoor Road, Meltham, Huddersfield, West Yorkshire HD7 3JY
Tel: 01484 854898
Fax: 01484 854899

Environment & Health News
www.ehn.clara.net

Specialities: Magazine; environment
Website: Publishers of a magazine which gives the latest news and research on the impact of environmental pollution, modern technology and lifestyle on our health, in short easy-to-read format with references. A free copy may be ordered on-line. A wide selection of articles on a variety of topics may be consulted on the site.
Links: Yes
Email: ehn@onetel.net.uk
Address:
The Environmental Health Trust,
Muir of Logie, Dunphail,
Forres IV36 2QG, Scotland
Tel: 01309 611200
Fax: 01309 611200

envirostore.co.uk
www.envirostore.co.uk

Specialities: Resource; directory
Website: On-line information point for the UK. Features a range of environmentally sound alternatives including, clothes, toys, furnishings and toiletries. Allows sourcing of products, services, organisations, initiatives, editorial and educational content. Site is organised under business and personal.
Email: info@envirostore.co.uk
Address: Greenmedia, Bermuda House, 3a Dinsdale Place, Sandyford, Newcastle-upon-Tyne, Tyne and Wear NE2 1BD
Tel: 0870 7893666

Enzafruit Worldwide
www.enzafruit-worldwide.co.uk

Speciality: Wholesaler
Website: Major fruit importer servicing multiple retailers. All year supply of apples, pears, kiwis, mangos and avocados. Website was under construction at time of writing.
Contact: Nick Rickett
Email:
nick.rickett@enzafruit-worldwide.co.uk
Address: West Marsh Road, Spalding, Lincolnshire PE11 2BB
Tel: 01775 717000
Fax: 01775 717001

Equal Exchange
www.equalexchange.co.uk

Specialities: Coffees; teas
Website: Innovative fairtrade organic company dealing in tea, coffee, honey, butters, confectionery, cocoa and sugar. Website not only gives product information but also producer details. There is information on stockists.
Links: Yes
Contact: Sam Roger
Email: info@equalexchange.co.uk
Address: 10a Queensferry Street, Edinburgh EH2 4PG, Scotland
Tel: 0131 220 3484
Fax: 0131 220 3565

Escential Botanicals Ltd
www.escential.com

Specialities: Skincare; essential oils
Website: Products contain only the purest and highest quality natural active ingredients that act gently on the skin. They are divided into body, face and hair. There is a page setting out the standards of the company.
On-line ordering: Yes
Credit cards: Most major credit cards
Email: archie@escential.com
Address: Unit 25, Mountbatten Road, Kennedy Way, Tiverton, Devon EX16 6SW
Tel: 01884 257 612
Fax: 01884 258 928

Essential Trading
www.essential-trading.co.uk

Specialities: Distributor; wholesaler; vegetarian & vegan
Website: UK based workers co-operative. Major UK distributor to independent businesses in the UK and abroad with a catalogue which includes over 5,000 lines, of which 50% are organic. Many unusual and pioneering product ranges brought together from the UK and abroad. It deals with organic farms in Italy, Egypt and New Zealand.
Links: Yes
Export: Yes
Catalogue: Yes
Contact: Mark Woollard
Email:
contact-us@essential-trading.co.uk
Address: Unit 3, Lodge Causeway Trading Estate, Fishponds, Bristol BS16 3JB
Tel: 0117 958 3550
Fax: 0117 958 3551

Ethical Consumer
www.ethicalconsumer.org

Specialities: Organisation; magazine; environment
Website: Alternative consumer organisation looking at the social and environmental records of the companies behind the brand names. It is also publisher of the magazine *Ethical Consumer*. Section headings on the site include magazine, research, boycotts, philosophy and corporate critic.
Email: mail@ethicalconsumer.org
Address: Unit 21, 41 Old Birley Street, Manchester M15 5RF
Tel: 0161 226 2929
Fax: 0161 226 6277

Ethical Financial Ltd
www.ethical-financial.co.uk

Specialities: Ethical investment; financial
Website: Large independent financial adviser specialising in socially responsible investment with franchise offices throughout the UK. Website was under construction at the time of writing.
Email: ethical@ethical-financial.co.uk
Address: Regus House, Malthouse Avenue, Cardiff Gate Business Park, Cardiff CF23 8RU, Wales
Tel: 029 2026 3622
Fax: 029 2026 3729

The Ethical Investors Group
www.oneworld.org/ethical-investors

Specialities: Financial advice; ethical investment
Website: A socially responsible UK based company offering independent advice exclusively on green, ethical, environment-friendly and socially responsible investment funds. Website gives information on ethical investment, funds, investors and investment companies. It gives details on services and provides an ethical fund directory with a rating code.
Email: info@ethicalinvestors.co.uk
Address: Greenfield House, Guiting Power, Cheltenham, Gloucestershire GL54 5TZ
Tel: 01451 850777

Ethical Junction
www.ethical-junction.org

Specialities: Organisation; ethical choices; finance
Website: This is a one-stop shop for ethical organisations and ethical trading. The site has an ethical shopping centre, a search engine that only searches ethical organisations, and the directory. Indexed alphabetically, by category and by region. It is full of hundreds of organisations representing a huge range of businesses and interests. The shop covers fairtrade goods, food and drink, books and publications and superstores.
Email: info@ethical-junction.org
Address: Fourways House, 16 Tarriff Street, Manchester M1 2FN
Tel: 0161 236 3637

Ethical Money Ltd
www.ethicalmoneyonline.com

Specialities: Ethical investment; financial
Website: Provides a one-stop shop for a wide range of ethical financial services. Website gives a summary of the services offered with definitions of ethical. Section headings on the website include ethical ISA guide, ethical fund selector, ethical portfolio management and discounts. There is an on-line facility for registering as a member. There is no charge for membership.
Email: info@ethicalmoneyonline.com
Address: 61a Friargate, Preston, Lancashire PR1 2XS
Tel: 01772 558 557
Fax: 01539 825 041

The Ethical Partnership
www.the-ethical-partnership.co.uk

Specialities: Ethical investment; financial services
Website: Specialist adviser on socially responsible and ethical investments. The website explains how they can help customers keep their values in harmony with financial planning. There are pages on the performance of ethical funds, independent financial advice, an ethical questionnaire and advice on how to proceed. The contact page gives details of partners throughout the UK.
Email: enquiries@the-ethical-partners.co.uk
Address: 53 Whyteleafe Hill, Whyteleafe, Surrey CR3 0AJ
Tel: 020 8763 1717
Fax: 020 8763 2716

Ethical Wares
www.ethicalwares.com

Specialities: Footwear; vegetarian & vegan
Website: Ethically based company run by vegans who seek to trade in a manner which does not exploit humans, animals or the wider environment. Catalogue includes vegan (non-leather) shoes, clothing and accessories of various kinds. Website gives illustrated on-line catalogue. There is a sizing conversion chart. Secure on-line ordering is available.
On-line ordering: Yes
Email: vegans@ethicalwares.com
Address: Caegwyn, Temple Bar, Felinfach, Ceredigion SA48 75A, Wales
Tel: 01570 471155
Fax: 01570 471166

Ethos
www.ethosbaby.com

Specialities: Nappies; babycare
Website: Fine natural baby products including reusable cotton nappies, natural baby toiletries, wooden toys, all-terrain buggies and other nursery items. All items can be ordered on-line or there is an email request form for a printed brochure.
Links: Yes
On-line ordering: Yes
Credit cards: Most major credit cards
Catalogue: Brochure
Email: enquiries@ethosbaby.com
Address: 37 Chandos Road, Bristol BS6 6PQ
Tel: 0117 907 3320
Fax: 0117 907 3320

European College of Bowen Studies (ECBS)
www.bowentechnique.com

Specialities: Bowen Technique; courses; practitioner training
Website: Website has the largest source of information on the Bowen technique, including research findings and a full list of therapists. There are also details of ECBS courses with dates and training format, venues and fees, as well as information on teachers. There is a suggested reading page with on-line ordering for the recommended titles through Amazon.co.uk.
Email: enquiries@thebowentechnique.com
Address: 38 Portway, Frome, Somerset BA11 1QU
Tel: 01373 461 873
Fax: 01373 461 873

The European Shiatsu School
www.shiatsu.org.uk

Specialities: Shiatsu; courses; practitioner training
Website: The website explains the principles of shiatsu and gives information on the school and its aims and philosophy. The essence of the philosophy is to perceive individual harmony as the key to social harmony. The school therefore not only offers competent shiatsu but also develops increased awareness and equilibrium. There are details of current course options and fees, practitioner training, post-graduate, teacher training and other courses.
Email: info@shiatsu.org.uk
Address: High Banks, Lockeridge, Marlborough, Wiltshire SN8 4EQ
Tel: 01672 513 444
Fax: 01672 861 459

D W & C M Evans
www.caerfai.co.uk

Specialities: Farm shop; box scheme; camping; cheese; cottages; holidays
Website: Organic dairy farm producing three types of cheese from unpasteurised milk. Holiday cottages and a campsite are available. The website gives a description of the farm and detailed, illustrated descriptions of the cottages. There is also information on the cheeses and cheese making.
Contact: Wyn Evans
Email: info@cheeseuk.com
Address: Caerfai Farm, St Davids, Haverfordwest, Pembrokeshire SA62 6QT, Wales
Tel: 01437 720548
Fax: 01437 720548

Eve Taylor Aromatherapy
www.eve-taylor.com

Specialities: Aromatherapy products; skincare; courses
Website: Offers a range of expert blended professional and retail facial oils, body oils and various accessories such as oil burners and environmental fragrancing kits. Details of products for men, women and lifestyle are given on the web pages. There is also information on aromatherapy training. There is an email form for enquiries about local stockists of the product range.
Links: Yes
Contact: Alan Taylor
Email: sales@eve-taylor.com
Address: Eve Taylor (London) Ltd, 8 - 9 Papyrus Road, Werrington, Cambridgeshire PE4 5BH
Tel: 01733 321101
Fax: 01733 321633

Excel Industries
www.warmcel.com

Specialities: Insulation; building materials
Website: Manufactures insulating materials made from recycled newspaper, softwood off-cuts and forest thinnings. The website explains the ecological approach and has pages on air-tightness, installation, performance, projects and approval.
Links: Yes
Email: excel.ind@btinternet.com
Address: 13 Rassau Industrial Estate, Ebbw Vale, Gwent NP3 5SD, Wales
Tel: 01495 350655
Fax: 01495 350146

Expressions Breastfeeding
www.centralmedical.co.uk

Specialities: Breast-feeding; maternity
Website: Company sell breast pumps and breast-feeding accessories. There is a full product list on the website with an on-line ordering facility. There is also a rental scheme with details given on the site. A useful information table will help in choosing the most appropriate product.
On-line ordering: Yes
Credit cards: Visa; Mastercard; Switch; Delta
Email: sales@centralmedical.co.uk
Address: CMS House, Basford Lane, Leekbrook, Leek, Derbyshire ST13 7DT
Tel: 01538 386 650
Fax: 01538 399 572

The Fairtrade Foundation
www.fairtrade.org.uk

Specialities: Organisation; fairtrade
Website: Foundation aims to guarantee a fair deal for poor producers. The website provides a definition of fairtrade. There is a product directory of retail and catering products. There is also information on individual producers. There is a facility to order promotional materials and subscribe to the newsletter. There is also a photo library. A new resource for teachers, *Fairtrade in the Classroom*, is available on-line.

Address: Suite 204,
16 Baldwin's Gardens,
London EC1N 7RJ
Tel: 020 7405 5942
Fax: 020 7405 5943

Faith Products Ltd
www.faithproducts.com

Specialities: Aromatherapy products; bodycare
Website: Some of the products are also organic and there has been no animal testing since 1985. Products are available through retail outlets although eventually it will be possible to order from the website. There is a listing of products on the web pages. An own label and contract manufacturing service is also offered and details are given on the site.
Contact: Aaron Rose
Email: sales@faithproducts.com
Address: Unit 5, Kay Street,
Bury, Lancashire BL9 6BU
Tel: 0161 764 2555
Fax: 0161 762 9129

Fargro Ltd
www.Fargro.co.uk

Specialities: Wholesaler; fertiliser; pest control; gardening
Website: Horticultural wholesale business supplying organic fertiliser and organic pest control products. Website provides company profile, product range, what's new, croptalk and newsletter.
Links: Yes
Contact: J McAlpine
Email: promos-fargro@btinternet.com
Address: Toddington Lane,
Littlehampton, West Sussex BN17 7PP
Tel: 01903 721591
Fax: 01903 730737

Farm-A-Round Ltd
www.farmaround.co.uk

Specialities: Vegetables; prepared vegetables; fruit; home delivery
Website: Specialises in prepared vegetables and fruit such as sliced onions, ready-to-cook carrot sticks and fruit salads. Works with farmers in the UK and the Mediterranean. Delivers across most of Greater London and also supplies hotels, shops and caterers. An information pack may be requested on-line. Website was under development at time of writing.
Catalogue: Information pack
Contact: Isobel Davies
Email: info@farmaround.co.uk
Address: (office)
B136 - B134 New Covent Garden Market, Nine Elms Lane,
London SW8
Tel: 020 7627 8066
Fax: 020 7627 4698

Farrow & Humphreys Ltd
www.fpisales.com

Specialities: Aromatherapy products; herbs; skincare
Website: Europe's largest aromatic ingredient supplier. Manufacturer of a range of body balms, soaps and bath products. Website under re-construction at the time of writing.
Catalogue: Free brochure available
Email: paul@fpisales.com
Address:
Meadow Park Industrial Estate,
Bourne Road, Essendine, Lincolnshire
PE9 4LT
Tel: 01780 482 200
Fax: 01780 482 112

Federation of City Farms & Community Gardens
www.farmgarden.org.uk

Specialities: Organisation; city farms; farming; agriculture
Website: Locally based projects working with people, animals and plants. Each one is unique. They aim to be community-led and managed, empowering those involved through a sustainable approach to what they do. Website lists the characteristics and benefits of the projects. It provides news and events, and details of how to start new projects. There will be a search facility to locate city farms and in the meanwhile there is a downloadable list of city farms open to the public.
Links: Yes
Email: admin@farmgarden.org.uk
Address: The Green House, Hereford Street, Bedminster, Bristol BS3 4NA
Tel: 0117 923 1800
Fax: 0117 923 1900

Federation of Holistic Therapists (FHT)
www.fht.org.uk

Specialities: Holistic therapy; organisation
Website: The UK's largest organisation representing beauty therapists, aromatherapists, reflexologists, holistic and complementary therapies, masseurs, electrologists, fitness and sports therapists, salon and clinic owners and therapy lecturers. Website sets out aims and objectives, explains who is represented, and gives information on membership benefits, treatments and liability insurance for members and latest news. There is an international register of complementary therapists alphabetically listed by country.
Email: info@fht.org.uk
Address: 3rd Floor, Eastleigh House, Upper Market Street, Eastleigh, Hampshire SO54 9FD
Tel: 023 8048 8900
Fax: 023 8048 8970

M Feller Son & Daughter Organic Butchers
www.mfeller.co.uk

Speciality: Meat;
Website: Business dedicated to organic British meat products. The shop is situated in the Oxford Covered Market and produces traditionally made organic sausages. Free local delivery. The meat price list is available on-line.
Email: mfeller@mfeller.co.uk
Address: 54-55 Oxford Covered Market, High Street & Cornmarket, Oxford, Oxfordshire
Tel: 01865 251164
Fax: 01865 200553

Feng Shui Catalogue
www.thefengshuicatalogue.co.uk

Speciality: Feng Shui
Website: The website provides an introduction to the principles of feng shui whose aim is to create a balanced and harmonious environment for all individuals. The website provides an on-line catalogue of feng shui related products for purchase as well as giving information on courses and exhibitions. Eventually this is planned to be a feng shui resource site.
On-line ordering: Yes
Credit cards: Most major credit cards
Catalogue: Yes
Contact: Georgina Burns
Email: thefengshhui.catalogue@virgin.net
Address: Green Dragon House, 16 Goldsmiths Road, London W3 6BH
Tel: 020 8992 6607
Fax: 020 8992 6607

Feng Shui Society
www.fengshuisociety.org.uk

Specialities: Feng shui; organisation
Website: A non-profit organisation formed to advance feng shui principles and concepts as a contribution to the creation of harmonious environments for

individuals and society in general. Website gives information on consultants, schools and teachers, events, regional groups and points of contact.
Email: info@fengshuisociety.org.uk
Address: 377 Edgware Road, London W2 1BT
Tel: 07050 289 200
Fax: 01423 712 869

Ferryman Polytunnels
www.ferryman.uk.com

Specialities: Polytunnels; gardening
Website: Over 400 different types of polytunnels manufactured for both professional and amateur gardeners. Website gives full details of dimensions, prices and specifications of a wide variety of polytunnels. It also covers accessories. There are illustrated instructions for erecting a polytunnel. A brochure can be downloaded or requested by email.
Export: Yes
Contact: Hugh Briant-Evans
Email: info@ferryman.uk.com
Address: Bridge Road, Lapford, Crediton, Devon EX17 6AE
Tel: 01363 83444
Fax: 01363 83050

Fertiplus Garden Products
www.fertiplus.co.uk

Specialities: Fertilisers; gardening
Website: Organic horticultural fertilisers including pelleted poultry manure. Suppliers to all areas of horticulture from mail order to growers. Information pack available. Website provides secure on-line ordering and details of the products. Full product information is given together with stockist information by area.
Links: Yes
On-line ordering: Yes
Credit cards: Most major cards
Catalogue: Information pack
Contact: James Holton
Email: jim@fertiplus.co.uk
Address: North Lodge, Orlingbury Road, Isham, Kettering, Northamptonshire NN14 1HW
Tel: 01536 722424
Fax: 01536 722424

Fibrowatt Ltd
www.fibrowatt.com

Speciality: Alternative energy
Website: Developer, builder and operator of electricity power stations using poultry litter as the fuel. Website explains the technology and environmental benefits of its power stations. There are pages on individual working power stations. It also gives information on overseas projects in the USA and the Netherlands.
Email: naomi.stevenson@fibrowatt.com
Address: Astley House, 33 Notting Hill Gate, London W11 3JQ
Tel: 020 7229 9252
Fax: 020 7221 8671

Fieldfare Organics
www.fieldfare-organics.com

Specialities: Box scheme; fruit; vegetables; meat; dairy; wholefoods
Website: Home delivery in Bedfordshire, Buckinghamshire, Hertfordshire and Middlesex of all organic requirements. On-line order form provides full list of organic products. There is also a wholefoods list and a meat list. A password is required for on-line ordering. Website provides information on the service, weekly news and weekly recipe.
Links: Yes
On-line ordering: Yes
Credit cards: Visa; Mastercard; Switch
Contact: Ray Calow
Email: office@fieldfare-organics.com
Address: Oakcroft, Dudswell Lane, Berkhamstead, Hertfordshire HP4 3TQ
Tel: 01442 877363
Fax: 01442 879950

Findhorn Flower Essences
www.findhornessences.com

Specialities: Essences; flower essences
Website: Established small family business producing a potent range of Scottish flower essences from the famous Findhorn Community gardens by the sun-infusion method pioneered by Dr Edward Bach. On the website there is information on how to use and choose the essences and how to make a personal assessment. Flower essences are listed and individually described. There is an on-line ordering service. A book on Findhorn Flower essences is available.
Links: Yes
On-line ordering: Yes
Contact: David Steel
Address: Wellspring, The Park, Findhorn Bay, Forres, Moray IV36 3TY, Scotland
Tel: 01309 690129
Fax: 01309 691300

Finlay's Foods
www.finlaysfoods.co.uk

Specialities: Caviar alternative; vegetarian; kosher
Website: Cavi'Art is the world's only vegetarian caviar alternative. Ambient, GMO-free and kosher certified. It is claimed to be virtually indistinguishable from the real thing but is one thirtieth of the price. There is an on-line order form.
Credit cards: Visa; Mastercard
Export: Yes
Contact: Neville Finlay
Email: finlay001@aol.com
Address: 60 Sherborne Street, Manchester M8 8LR
Tel: 0161 833 3303
Fax: 0161 839 0600

Fired Earth
www.firedearth.com

Speciality: Flooring
Website: Natural floorings including stone and terracotta and also rugs. There are showrooms throughout UK. The company provides a source of ideas and inspirations. Website gives selected product listing of paints, rugs, natural flooring, fabrics, floor tiles and wall tiles. Showrooms in the UK are locatable on the map.
Catalogue: Yes
Email: enquiries@firedearth.com
Address: Twyford Mill, Oxford Road, Adderbury, Oxfordshire OX17 3HP
Tel: 01295 814315
Fax: 01295 810832

First Foods
www.first-foods.com

Specialities: Ice cream; non-dairy
Website: Range of non-dairy products including unique ice creams full of the goodness of oats with the taste of traditional ice cream. The website gives general information on the products and brands. There is also a page on First Milk, a cereal-based alternative to dairy and soya milk.
Contact: Jim Evans
Email: innovations@first-foods.com
Address: PO Box 140, Amersham, Buckinghamshire HP6 6XD
Tel: 01494 431355
Fax: 01494 431366

First Quality Foods
www.firstqualityfoods.co.uk

Specialities: Pasta; confectionery; vegetarian & vegan
Website: Manufacturers of the Sammy brand of couscous and polenta - offering a wide selection of plain, flavoured and organic products. Also producers of organic bars in 16 fruit and nut flavours that are vegetarian, vegan and wheat-free. General product information is given on the website.
Contact: Trudy Jenkins
Email: info@firstqualityfoods.co.uk
Address: Unit 29, The Beeches, Yate, Bristol BS37 5QX
Tel: 01454 880044
Fax: 01454 853355

Flow Forms
www.flowforms.com

Speciality: Water sculptures
Website: Unusual water sculptures specially designed and following certain principles related to holistic theory. The ideas revolve around the texturing of water through basins carefully designed to produce a double vortex. Website has headings on design ideas, gallery of designs, education, installation, exterior and interior projects and solar power. There is also a price guide.
Links: Yes
Email: info@flowforms.com
Address: c/o Docklands Garden Centre, 244-246 Ratcliffe Lane, London E14 7JE
Tel: 020 7923 9622
Fax: 020 7790 5025

Food Certification (Scotland) Ltd
www.foodcertificationscotland.co.uk

Specialities: Organisation; certification
Website: Independent third party certification body committed to encouraging the production of high quality fishery, food and agricultural products. The website explains how it operates and what the certification schemes are.
Address: Redwood,
19 Cuduthel Road, Inverness,
Inverness-shire IV2 4AA, Scotland
Tel: 01463 222251
Fax: 01463 711408

The Food Commission
www.foodcomm.org.uk

Specialities: Organisation; nutrition
Website: Campaigns for safer, healthier food in the UK. Publishes the *Food Magazine* which reports on genetic engineering, pesticides, additives, food irradiation, food labelling and animal welfare. The website provides information on the commission, books, reports, the *Food Magazine*, press releases and latest news.
Email: foodcomm@compuserve.com
Address: 94 White Lion Street,
London N1 9PF
Tel: 020 7837 2250
Fax: 020 7837 1141

Food Revolution Ltd
www.foodrevolution.com

Specialities: Groceries; home delivery
Website: Committed to drawing together the best foods from Britain and beyond, delivered to the door. The website includes section headings for suppliers, products, meal ideas and the club. As an ethical company suppliers get a fair deal and most of the foods are organic and free range.
On-line ordering: Yes
Credit cards: Visa; Mastercard; Switch
Email: sales@foodrevolution.com
Address: 1 Mead House,
Little Mead Industrial Estate,
Alford Road, Cranley, Surrey GU6 8ND
Tel: 01483 277 082
Fax: 01483 277 082

Forbo-Nairn Ltd
www.forbo-nairn.co.uk

Specialities: Flooring; linoleum
Website: Manufacturer of linoleum, a product made from natural materials. The website gives help in selection and has a sample request page.
Email: headoffice@forbo-nairn.co.uk
Address: PO Box 1, Den Road,
Kirkcaldy, Fife KY1 2SB, Scotland
Tel: 01592 643777
Fax: 01592 643999

Forest Products
www.forestproducts.co.uk

Specialities: Preserves; chutneys; mustards; jams
Website: Manufacturers of organic, preserves, chutneys and mustards. There is an on-line brochure listing the products and there is also an email request form for a printed brochure. On-line ordering is being developed - in the meantime orders may be phoned, faxed or emailed. Direct ordering is geared to trade customers.
Catalogue: Brochure
Contact: Roger Stagg
Email: mail@forestproducts.co.uk
Address:
24 - 28 Dreadnought Trading Estate, Bridport, Dorset DT6 5BU
Tel: 01308 458111
Fax: 01308 420900

Forever Living Products
www.aloevera.co.uk

Speciality: Aloe vera products
Website: The website provides information on the business opportunity offered by becoming part of the FLP distributor marketing network. The principles of network marketing are analysed and the scope and function of the marketing plan explained. The website also exists to provide information on the properties of aloe vera and products based on it as well as to provide an independent research resource on the effect of aloe vera in humans and animals. Products can be ordered on-line.
On-line ordering: Yes
Credit cards: Visa; Mastercard; Switch; Delta; Amex
Email: orders@aloevera.co.uk
Address: 16 Treverbyn Road, Truro, Cornwall TR1 1RG
Tel: 01872 276398
Fax: 01872 276 398

Forever Young International
www.forever-young-health.com

Speciality: Food supplement
Website: Immunity food formula with live sprouted broccoli seeds. The product contains 12 key superfoods which synergistically combine to support the immune organs (thymus, spleen, lymph and blood) and strengthen immunity. The company supplies retail outlets and exports to Europe and further afield. There is a product listing on the website for direct on-line ordering.
On-line ordering: Yes
Credit cards: Most major cards
Export: Yes
Contact: Paul Dracup
Email: forever-young@lineone.net
Address: 7 Falcon Court, St Martins Way, Wimbledon, London SW17 0JH
Tel: 020 8944 7442
Fax: 020 8946 3572

Fragrant Earth
www.fragrant-earth.com

Specialities: Aromatherapy products; herbal remedies; homoeopathy; bodycare
Website: The company is active in the manufacture, production and distribution of raw materials for the phytopharmaceutical, cosmetics, toiletries, aromatherapy and other industries that require raw materials of a high standard. The list of clinical and speciality material is designed for practitioners and professionals. Website gives select details of the printed catalogue which is available on request. It also provides definitions of essential oils and vegetable oils.
Catalogue: Yes
Contact: Debbie Seaton
Email: all-enquiries@fragrant-earth.com
Address: Orchard Court, 3A Magdalene Street, Glastonbury, Somerset BA6 9EW
Tel: 01458 831216
Fax: 01458 831361

FRAGRANT STUDIES INTERNATIONAL LTD
www.fragrantstudies.com

Specialities: Vibrational healing; aromatherapy; courses
Website: Promoting aromatherapy education for all with courses for qualified and training aromatherapists held at Glastonbury. Many classes are modular and are accredited for continuous education programmes. Website provides a courses schedule, information on the tutors and on Glastonbury and surroundings, and an on-line registration form.
Email:
fragrant-studies@fragrant-earth.com
Address: Orchard Court,
3A Magdalene Street, Glastonbury,
Somerset BA6 9EW
Tel: 01458 835920
Fax: 01458 831361

FREEPLAY ENERGY LTD
www.freeplay.net

Specialities:
Self-powered energy products; alternative energy
Website: Pioneers of technology to serve the needs of humankind though the self-sufficient electronics industry. Their first product was the wind-up radio. The range, which now includes flashlights, is described and illustrated on the website with a 360 degree viewing option. There is a search facility for a local retailer and a links facility for on-line connection to e-tailers.
Email: info@freeplay.co.uk
Address: Cirencester Business Park,
Love Lane, Cirencester,
Gloucestershire GL7 1XD
Tel: 01285 659 559
Fax: 01285 659 550

FREERANGERS
www.freerangers.co.uk

Specialities: Footwear; fleeces; vegan
Website: Animal free products approved by the Vegan society. The range covers footwear and fleeces for children, men and women as well as accessories such as scarves and rucksacks.
Links: Yes
On-line ordering: Yes
Credit cards: Most major credit cards
Catalogue: Yes
Address:
9b Marquis Court, Low Prudhoe,
Northumberland NE42 6PJ
Tel: 01661 831 781
Fax: 01661 830 317

FRESH & WILD
www.freshandwild.com

Specialities: Supermarket; delicatessen; juice bar; herbal remedies
Website: Organic foods and natural remedies supermarket run as community grocers. On-line there is a listing of their community stores with addresses in the London area. There is also a mission statement on their environmental policy and a recipe page.
On-line ordering: No
Email: shop@freshandwild.com
Address: 49 Parkway,
London NW1
Tel: 020 7428 7575

THE FRESH FOOD COMPANY
www.freshfood.co.uk

Specialities: Supermarket; fruit; vegetables; box schemes; meat; household products
Website: Over 4,500 organic and wild-harvested foods, wines and household necessities available on-line or by phone, fax or email. Delivery once a week nationwide. There are on-line recipes and product research information with quick search facilities.
Links: None
On-line ordering: Yes
Credit cards:
All major credit & debit cards
Catalogue: Weekly email newsletter
Email: organics@freshfoods.co.uk
Address: 326 Portobello Road,
London W10 5RU
Tel: 020 8969 0351
Fax: 020 8964 8050

Fresh Network
www.fresh-network.com

Specialities: Raw food; nutrition; organisation
Website: The Fresh Network brings together those who take an interest in a 100% or high percentage raw food diet. New website was under construction at the time of writing. Current website contains brochure discussing the concept of freshness and explaining what membership of the network entails. It gives details of events and a full mail order catalogue (books, juicers, dehydrators, sprouting equipment etc.)
Links: Yes
Email: info@fresh-network.com
Address: PO Box 71, Ely, Cambridgeshire CB7 4GU
Tel: 0870 800 7071
Fax: 0870 800 7070

Fresh Water Filter Company
www.freshwaterfilter.com

Speciality: Water purifiers
Website: The aim of the company is to supply high quality products and services and to produce the finest, healthy water that creates human, environmental and commercial benefits. Products listed on the website include ceramic, gravity, and shower filters, whole-house set ups, softeners, distillers, coolers, custom units and refills. Full specifications are provided and there is an on-line order form.
On-line ordering: Yes
Credit cards: Most major credit cards
Email: mail@freshwaterfilter.com
Address: Gem House, 895 High Road, Chadwell Heath RM6 4HL
Tel: 020 8597 3223
Fax: 020 8590 3229

Friends of the Earth
www.foe.co.uk

Specialities: Organisation; environment
Website: Leading UK pressure group campaigning on a broad range of local, national and international environmental issues. Extensive website covers wide-ranging topics including a why real food is important section which gives details of campaigns on GMOs and pesticides and provides in-depth information in the form of documents on related subjects. There are details of many other environmental campaigns on the site. Membership details are available.
Email: info@foe.co.uk
Address: 26 - 28 Underwood Street, London N1 7JQ
Tel: 020 7490 1555
Fax: 020 7490 0881

Friends of the Western Buddhist Order
www.lbc.org.uk

Specialities: Buddhism; meditation
Website: Website provides information on courses on Buddhism and meditation including on-line teaching via a link to a US meditation teacher. There are details and dates of retreats and events, together with a calendar.
Email: info@lbc.org.uk
Address: London Buddhist Centre, 51 Roman Road, London E2 0HU
Tel: 020 8981 1225
Fax: 020 8980 1960

Full Moon Communications
www.naturalproducts.co.uk

Specialities: Show organisers; magazines
Website: Organisers of Natural Products Europe, The Organic Living Show and other events for the natural and organics product sector. Publishers of *Natural Products News*, the UK's leading trade magazine for the industry. Website gives details of the shows, registering and booking information.
Contact: Robin Bines
Email: rbines@naturalproducts.co.uk
Address: 58 High Street, Steyning, West Sussex BN44 3RD
Tel: 01903 817300
Fax: 01903 817310

Futura Foods UK Limited
www.nordex-food.co.uk

Speciality: Dairy
Website: A subsidiary of Nordex Food, a Danish manufacturing and dairy trading company, marketing organic dairy products in the UK in the retail and wholesale sectors. Website gives a company introduction and a listing of products available to retailers, wholesalers, distributors and the catering trade,
Contact: Claire Jackson
Email: dairyland@nordex-food.co.uk
Address: Wynchfield House, Kingscote, Nr Tetbury, Gloucestershire GL8 8YN
Tel: 01666 890500
Fax: 01666 890522

G's Marketing
www.gs-marketing.com

Specialities: Wholesaler; vegetables
Website: Wholesale suppliers of organic vegetables and salads to multiple outlets and to independent retailers and caterers. Website gives history and aims of this family owned company.
Contact: Rowen Markie
Email: info@gs-marketing.com
Address: Barway, Ely, Cambridgeshire CB7 5TZ
Tel: 01353 727200
Fax: 01353 723021

Gaia Distribution
www.gaiadistribution.com

Speciality: Games
Website: Importers of ecological games and green goods. Website provides details of the games, age levels, number of players for each game and describes the game ideas.
Links: Yes
Catalogue: Yes
Contact: Joe & Carme Brunner
Email: mail@gaiadistribution.com
Address: 79 Mackie Avenue, Hassocks, Sussex BN6 8JN
Tel: 01273 843 503
Fax: 01273 843 503

Gale & Snowden
www.ecodesign.co.uk

Speciality: Architects
Website: Ecological and energy efficient architects. Website gives information on the practice philosophy, research and services. There is also a practical profile and examples of selected projects ranging from a turf-roof house to a sustainable village. There are examples of eco-renovation.
Links: Yes
Email: galesnow@ecodesign.co.uk
Address: 18 Market Place, Bideford, Devon EX39 2DR
Tel: 01237 474 952
Fax: 01237 425 669

Ganesha
www.ganesha.co.uk

Specialities: Furnishings; clothing; Indian goods
Website: Importers of quality home furnishings and artifacts from India, sourced from co-operatives and producer associations thus supporting an alternative vision of trade by extending markets for marginal producers and supplying the UK with life-enhancing goods. Website gives social and environmental policy, and details of producer groups. There is an on-line catalogue showing scarves, wraps, shawls, cushions, bedcovers, artifacts, carpets and clothing.
Links: Yes
On-line ordering: Yes
Credit cards: All major credit cards
Contact: Jo Lawbuary & Purnendy Roy
Email: ggmail@ganesha.co.uk
Address: 3 Gabriel's Wharf, 56 Upper Ground, London SE1 9PP
Tel: 020 7928 3444
Fax: 020 7928 3444

Gardentrouble
www.gardentrouble.co.uk

Speciality: Gardening information
Website: Garden trouble aims to answer specific gardening queries or problems. If the solution isn't on the database you will be asked to email your question. There is also a directory of garden products and services. There is a section or organic fruit and vegetables, covering fruit trees, rotation, avoiding pests, composting, and companion planting.
Links: Yes
Email: website@gardentrouble.co.uk

Garlic Genius
www.garlicgenius.co.uk

Speciality: Garlic
Website: Website provides an illustrated overview of the different types of garlic. There is also detailed coverage of the specially designed garlic cutter which is available. Full product list includes Spanish garlic, smoked garlic, elephant garlic, garlic strings, pickled garlic, dried garlic, garlic baker, garlic timer and recipe book. Orders can be placed by phone or fax on a printable order form
Credit cards: Visa; Mastercard
Email:
customerservices@garlicgenius.co.uk
Address: Paragon Promotions,
22 Brook Court, 81 Brook Road South,
Brentford, Middlesex TW8 0NY
Tel: 020 8847 3170
Fax: 020 8847 3170

General Council and Register of Naturopaths
www.naturopathy.org.uk

Specialities: Naturopathy; organisation
Website: Independent registering body establishing and maintaining standards, keeping a register and supervising the ethical behaviour of practitioners. Website gives information on the Council and its work, on professional conduct, training in naturopathy, in joining and in locating members.
Links: Yes
Email: admin@naturopathy.org.uk
Address: Goswell House,
2 Goswell Road, Street,
Somerset BA16 0JG
Tel: 01458 840 072
Fax: 01458 840 075

General Osteopathic Council
www.osteopathy.org.uk

Specialities: Organisation; osteopathy
Website: The website gives details of the council and its work. It also covers questions on osteopathy, fact sheets, osteopathy and medicine, careers information, and information for employers. There are pages on research and promotional materials can be ordered. There is also a search facility to locate registered osteopaths.
Email: info@osteopathy.org.uk
Address: Osteopathy House,
176 Tower Bridge Road,
London SE1 3LU
Tel: 020 7357 6655
Fax: 020 7357 0011

Genetic Engineering Network
www.geneticsaction.org.uk

Speciality: Genetic engineering
Website: Website was under construction at time of writing but gives information on GEN, local groups, getting active and safety protocol. GEN is a network of groups exchanging information and includes environmental organisations, retailers in the natural food trade, organic farmers and numerous local groups and individuals who oppose genetic engineering.
Email: genetics@gn.apc.org
Address: PO Box 9656,
London N4 4JY
Tel: 020 7690 0626

The Genetics Forum
www.geneticsforum.org.uk

Specialities: Organisation; genetic engineering
Website: A charity concerned with issues of genetic engineering, it is the UK's only public interest group devoted to policy development, campaigns and publications from a social, ethical and environmental viewpoint. The website provides a directory of national and international official and non-governmental bodies. The site also provides information on publications, articles, latest news, activities and history. There is an on-line index.
Links: Yes
Contact: Mark Raby
Email: geneticsforum@gn.apc.org
Address: 94 White Lion Street, London N1 9PF
Tel: 020 7837 9229
Fax: 020 7837 1141

Genewatch UK
www.genewatch.org

Specialities: Organisation; genetic engineering
Website: Independent policy research group concerned with the ethics and risks of genetic engineering. Briefings available on subscription. There is an on-line comprehensive and detailed database of GM crops and foods. Also available are topical news, articles, fact files, UK crop trials and press releases.
Links: Yes
Contact: Dr Sue Mayer
Email: mail@genewatch.org
Address: The Courtyard, Whitecross Road, Tideswell, Buxton, Derbyshire SK17 8NG
Tel: 01298 871898
Fax: 01298 872531

Get Real Organic Foods
www.get-real.co.uk

Specialities: Pies; desserts; Christmas puddings; vegetarian & vegan
Website: Producers of organic frozen pies: both meat and vegetarian and vegan. Also Christmas puddings, cakes and mince pies. Produce is sold to health shops and Waitrose. Website provides full listing of stockists.
Contact: S Beckett
Email: info@get-real.co.uk
Address: Shotton Farm, Harmer Hill, Shrewsbury, Shropshire SY4 3DN
Tel: 01939 210925
Fax: 01939 210925

A & D Gielty
www.aanddgielty.co.uk

Specialities: Vegetables; box scheme; farm shop
Website: Organic vegetables, salads and potatoes available throughout the year. One of the largest organic vegetable growers in the UK. Customers range from the general public to the largest supermarket chains. Website gives general company information and photographs.
Contact: Alf Gielty
Email: office@aanddgielty.co.uk
Address: Lyncroft Farm, Butchers Lane, Aughton Green, Ormskirk, Lancashire L39 6SY
Tel: 01695 421 712
Fax: 01695 422 117

Gilbert's
www.smoothhound.co.uk

Specialities: Bed & breakfast; holidays
Website: Gilbert's offers bed and breakfast using organic ingredients as much as possible. Website is part of a general accommodation site but gives prices, description and photograph.
Contact: Jenny Beer
Email: jenny@gilbertsbb.demon.co.uk
Address: Gilbert's Lane, Brookthorpe, Nr Gloucester, Gloucestershire GL4 0UH
Tel: 01452 812364
Fax: 01452 812364

Glenrannoch Vegetarian Guesthouse
www.glenrannoch.co.uk

Specialities: Guesthouse; holidays; vegetarian
Website: Simple website gives details of the tariff at this guest house which is 100% vegetarian and largely organic. A sample dinner menu is shown.
Links: Yes
Credit cards: Visa; Mastercard
Contact: Richard & Margaret Legate
Email: info@glenrannoch.co.uk
Address: Kinloch Rannoch, Nr Pitlochry, Perthshire PH16 5QA, Scotland
Tel: 01882 632 307

Global Eco
www.soapnut.com

Specialities: Bodycare; cleaning agents
Website: Multi-use organic washing product in powder form which has traditionally been used for centuries in India and which can be used on the body or for laundry. Website gives trade price list and general descriptions and details of the product.
Email: global@soapnut.com
Address: PO Box 2000, Bury St Edmunds, Suffolk IP33 9NL
Tel: 01284 700 170
Fax: 01284 700 270

The Global Retreat Centre
www.globalretreatcentre.com

Specialities: Retreats; meditation
Website: The centre is staffed and managed by teachers of meditation. It offers regular retreats, seminars, workshops and courses to learn meditation and self-management, develop personal skill and explore common spiritual values essential to the restoration of harmony within ourselves and the world. It is administered by the Brahma Kumaris World Spiritual University. The website provides background information and an events guide.
Email: info@globalretreatcentre.com
Address: Nuneham Park, Nuneham Courtenay, Oxford, Oxfordshire OX44 9PG
Tel: 01865 343 551
Fax: 01865 343 576

Go Organic
www.goorganic.co.uk

Specialities: Ready meals; sauces; soups; curries; vegetarian
Website: Soups made only from pure organic ingredients. No GMOs, artificial thickeners, artificial flavours, colours, hype or compromise. Products are available throughout the UK in natural and healthfood stores and in some prestige fine food retail outlets. Mail order is available with home delivery through www.freshfood.co.uk, telephone 0208 969 0351. A list of products with ingredients is given on the site.
Contact: Sheila Ross; Charlotte Mitchell
Email: email@goorganic.co.uk
Address: 24 Boswall Road, Edinburgh EH5 3RN Scotland
Tel: 0131 552 2706
Fax: 0131 552 2706

Golland Farm Garden Project
www.marketsquare.co.uk /chumleigh/bridleway

Specialities: Holidays; holiday cottages
Website: Holiday cottages on small mixed organic farm with fishing available. The website gives details of the cottages with photographs and detailed descriptions. Horticultural therapy is available for people with enduring mental health problems. There is also a vegetable bix scheme and community supported agriculture.

Contact: Jon Lincoln-Gordon
Email: golland@btinternet.com
Address: Golland Farm, Burrington, Umberleigh, Devon EX7 9JP
Tel: 01769 520263
Fax: 01769 520263

Good Food Distributors
www.goodfooddistributors.co.uk

Speciality: Wholesalers
Website: Wholesale delivery service of wholefood and healthfood products in Wales and neighbouring counties (map on website). Over 4,000 product lines supplied to healthfood outlets, wholefood shops and supermarkets. The product range is listed on the website. There is also an account application form.
Links: Yes
Contact: K Powell
Email: gfd.wholesale@btinternet.com
Address:
35 Ddole Road Industrial Estate,
Llandrindod Wells,
Powys LD1 6DF, Wales
Tel: 01597 824720
Fax: 01597 824760

Goodlife Foods
www.goodlife.co.uk

Specialities: Frozen vegetarian foods; vegetarian
Website: Leading specialist manufacturer of frozen vegetarian foods. All are GM free and are also increasingly organic. The website gives a company profile and lists the products and the organic range with illustrations.
Links: Yes
Email: enquiry@goodlife.co.uk
Address: 34 Tatton Court,
Kingsland, Grange, Warrington, Cheshire WA1 4FF
Tel: 01952 837810
Fax: 01952 838648

Goodman's Geese
www.goodmansgeese.co.uk

Specialities: Geese; turkeys
Website: Traditional free range geese and bronze turkeys which may be ordered via the website for collection or despatch from October to the end of December. Website provides pages of advice on buying, preparing and cooking geese together with recipes. There is a location map with road directions.
Contact: Judy Goodman
Email: info@goodmansgeese.co.uk
Address: Walsgrove Farm, Great Witley, Worcester, Worcestershire WR6 6JJ
Tel: 01299 896272
Fax: 01299 896889

GoodnessDirect
www.goodnessdirect.co.uk

Specialities: Special diets;
food supplements; sports nutrition
Website: 1,000 plus health products and items selected for special dietary needs. Search facility for products that are dairy-free, gluten-free, wheat-free, yeast-free and low fat. Also search facility by allergy or health requirement. There is a wholesale department for trade customers. Articles and information are included on the site.
Links: Yes
On-line ordering: Yes
Credit cards: Most major credit cards
Email: info@goodnessdirect.co.uk
Address: South March, Daventry, Northamptonshire NN11 4PH
Tel: 0871 871 6611
Fax: 01327 301135

Gordon Jopling Food Ingredients
www.jopling.co.uk

Specialities: Food ingredients; distributor
Website: Distributor and manufacturer of specialist food ingredients to food manufacturers, providing ambient stable, chilled and frozen ingredients to niche markets. Website gives company information and product listing. There is a newspage giving details of new products.
Contact: Cassie Marsden
Email: info@jopling.co.uk
Address: Shawfield Road, Carlton Industrial Estate, Carlton, Barnsley, South Yorkshire S71 3HS
Tel: 01226 733288
Fax: 01226 733113

Gossypium.co.uk
www.gossypium.co.uk

Specialities: Clothing; textiles; baby clothes; home textiles
Website: The eco-cotton store where everything is simple, organic and ethical. The website has section headings for casual clothing, baby and children, home textiles and gifts. Full information is provided including fabric details, size chart and washing instructions. Website also gives information on the company, its history and investment.
On-line ordering: Yes
Catalogue: Yes
Email: info@gossypium.co.uk
Address: 2/3 St Andrews Place, Lewes, Sussex BN7 1UP
Tel: 01273 897 509
Fax: 01273 897 507

Grafton International
www.graftons.co.uk

Specialities: Bodycare; toiletries; distributors; dental products
Website: UK distributors for Kneipp Werke, the German pharmaceutical company with 100 years experience of developing and manufacturing exclusively plant derived products. Web pages provide information on stockists both in and out of the UK. Email ordering is available to trade partners only.
Contact: Marie Lockwood
Email: sales@graftons.co.uk
Address: Birchbrook Park, Shenstone, Staffordshire WS14 0DJ
Tel: 01543 480100
Fax: 01543 480201

Graianfryn Vegetarian Guest House
www.vegwales.co.uk

Specialities: Guesthouse; holidays; vegetarian & vegan
Website: Vegetarian & vegan guesthouse in fine coastal and mountain scenery. Advance booking is essential. Website gives information on the food, rooms and prices, local attractions and activities.
Contact: Christine Slater
Email: info@vegwales.co.uk
Address: Penisarwaun, Caernarfon, Gwynedd LL55 3NH, Wales
Tel: 01286 871 007

Graig Farm Organics
www.graigfarm.co.uk

Specialities: Meat; fruit; vegetables; wines; dairy products; fish
Website: Mail order shop is arranged in departments: the butchery, the fishmongers, the pie shop, the dairy, groceries, the bakery, the greengrocery, non foods, the bookshop and the off-license. There are information pages including details of animal rearing and definitions of organic as well as detailed descriptions of Graig Farm. A large and well arranged site which expresses forthright opinions.
On-line ordering: Yes
Contact: Bob & Carolyn Kennard
Email: shop@graigfarm.co.uk
Address: Dolau, Llandidrod Wells, Powys LD1 5TL, Wales
Tel: 01597 851655
Fax: 01597 851991

The Granville Hotel
www.granvillehotel.co.uk

Specialities: Hotel; restaurant; holidays; vegetarian & vegan
Website: Organic produce served wherever possible in this environmentally friendly establishment. Also features Trog's Restaurant and Café Bar which is an organic vegetarian restaurant serving organic wines. The website describes the rooms, illustrating a selection, the restaurant and bar. Prices and booking information are supplied for email booking.
Contact: M Paskins
Email: granville@brighton.co.uk
Address: 124 King's Road, Brighton, East Sussex BN1 2FA
Tel: 01273 326302
Fax: 01273 728294

The Great Organic Picnic
www.thepicnic.org

Speciality: Events
Website: Annual organic picnic in London park with events, activities and entertainment, Website provides information, press releases and articles, a notice board and pictures of the event. A mailing list of interested supporters is being compiled.
Email: circletheory@compuserve.com
Address: 2 Old Brompton Road #205, London SW7 3DQ
Tel: 020 7865 8299
Fax: 020 7584 0369

Green and Black's
www.greenandblacks.com

Speciality: Chocolate
Website: The organic chocolate brand of Whole Earth Foods. There are pages of information on the history of chocolate, rainforests and the making of chocolate. There is a full product listing.
Links: Yes
Email: enquiries@wholeearthfoods.co.uk
Address: 2 Valentine Place, London SE1 8QH
Tel: 020 7633 5900
Fax: 020 7633 5901

Green and Organic
www.greenandorganic.co.uk

Specialities: Meat; fish; fruit; dairy; baby foods; home delivery
Website: All products are 100% organic guaranteed. Home delivery service in Hampshire, Surrey and surroundings. It is necessary to first obtain a product brochure and orders can then be emailed. Products are key coded for allergies and special dietary requirements. Over 10,000 products are available and there is also a wholesale division.
Links: Yes
Catalogue: Product brochure
Contact: Stuart & Tara Tilley
Email: info@greenandorganic.co.uk
Address: Unit 2, Blacknest Industrial Estate, Blacknest, Nr Alton, Hampshire GU34 4PX
Tel: 01420 520838
Fax: 01420 23985

Green Ark Animal Nutrition
www.greenark.mcmail.com

Specialities: Pet food; animal care
Website: Organic animal wholefoods and herbs by mail order. Small family-run business supplying unique organic whole grain and cereal mix for dogs and a range of herbal supplements to complete the diet and concentrate on problem areas of health. Website provides details of products, a feeding guide, diet sheet and price list.
Credit cards: Most major credit cards
Contact: Anne & Richard Langley
Email: greenark@cwcom.net
Address: Unit 7B, Lineholme Mill, Burnley Road, Todmorden, West Yorkshire OL14 7DH
Tel: 01282 606 810
Fax: 01282 606 810

Green Baby
www.greenbabyco.com

Specialities: Baby clothing; baby toiletries; toys; baby bedding; nappies; maternity products
Website: One-stop baby shop for natural products. A mail order service for planet-friendly parents providing natural and organic baby and maternity products. On-line ordering and home delivery.
On-line ordering: Yes
Credit cards: Most major cards
Contact: Jill Barker
Email: enquiries@greenbabyco.com
Address: 345 Upper Street, Islington, London N1 3QP
Tel: 020 7226 4345
Fax: 020 7226 9244

Green Board Game Company
www.greenboardgames.com

Speciality: Games
Website: Board games, card games and CD Rom's which are produced in an environmentally conscious way.
Catalogue: Yes
Email: info@greenboardgames.com
Address: 34 Amersham Hill Drive, High Wycombe, Buckinghamshire HP13 6QY
Tel: 01494 538 999
Fax: 01494 538 646

Green Books
www.greenbooks.co.uk

Specialities: Directories; books; publishers
Website: Publishers of the *Organic Directory*, which is now also on-line, in association with the Soil Association. Also other books on organic and related topics. Website gives catalogue listing of available titles by subject category together with book descriptions and reviews. Orders may be placed by email, phone or fax.
Credit cards: Visa; Mastercard; Access
Email: greenbooks@gn.apc.org
Address: Dartington, Totnes, Devon TQ9 6EB
Tel: 01803 863260
Fax: 01803 863843

Green Futures
www.greenfutures.org.uk

Speciality: Magazine; environment; sustainable development
Website: UK magazine on environmental solutions and sustainable futures which is available both in print and on the web. There is on-line searchable access to all recent issues of *Green Futures*. There are section headings on business and finance, cities and settlements, countryside, education, food, materials and recycling, policy and transport and travel. There is a listing of key organisations and business active in sustainable development with full contact details.
Email: post@greenfutures.org.uk
Address: Unit 55, 50 - 56 Wharf Road, London N1 7SF
Tel: 020 7608 2332
Fax: 020 7608 2333

Green Gardener
www.greengardener.co.uk

Specialities: Pest control; composting; seeds; gardening
Website: Family-run mail order company. Suppliers of natural pest controls such as nematodes and slug traps by mail order. They also supply organic vegetable seeds, compost converters, wormeries and worm composting and wildlife homes and feeders. There is a secure order for.
On-line ordering: Yes
Credit cards: Most major credit cards
Contact: Jonathan Manners
Email: jon@greengardener.co.uk
Address: 2a Norwich Road, Claydon, Ipswich, Suffolk IP6 ODF
Tel: 01473 833031

GREEN GUIDES
www.greenguideonline.com

Specialities: Directories; magazine; publishers
Website: Comprehensive eco-directory published in 9 regional editions. Very extensive on-line databases on green matters and health, searchable free of charge by a number of different categories. Also publishers of the magazine *Pure Modern Lifestyle*. Website gives information on directories currently in print and available. Books and magazine subscriptions may be ordered on-line.
On-line ordering: Yes
Email: info@greenguide.co.uk
Address: Green Guide Publishing, 271 Upper Street, London N1 2UQ
Tel: 020 7354 2709
Fax: 020 7226 1311

THE GREEN NETWORK
www.green-network.organics.co.uk

Specialities: Organisation; environment
Website: Dedicated to obtaining and passing on information on environmental pollution and health and address the rising public concern about health issues by providing independent scientifically based information to assist people in assessing the risks. Site provides information on a variety of topics.
Contact: Vera Chainey
Email: 01007273110@compuserve.com
Address: 9 Clairmont Road, Lexden, Colchester, Essex CO3 5BE
Tel: 01206 546902
Fax: 01206 766005

GREEN PEOPLE
www.greenpeople.co.uk

Specialities: Herbal tonics; skincare; haircare; suncare; health care; vegan
Website: Suppliers of organic dietary health care, suncare and personal care products. All organic formulations are based on natural plant extracts, vegetable oils, certified essential organic oils and herbs. Free from petro-chemicals, animal ingredients, synthetic additives and GMOs. Products may be bought on-line, by mail order and through health stores.
On-line ordering: Yes
Contact: Sue Losson
Email: organic@greenpeople.co.uk
Address: Brighton Road,
Handcross,
West Sussex RH17 6BZ
Tel: 01444 401 444
Fax: 01444 401 011

GREEN SCIENCE CONTROLS
www.green-science.com

Specialities: Pest control; gardening
Website: Green Science controls pests cleanly, effectively and without chemicals. Natural products are used for the protection of food both in the field as well as during processing and storage. The on-line catalogue is illustrated and allows secure electronic ordering. Free delivery within the UK.
On-line ordering: Yes
Credit cards: Visa; Mastercard; Switch
Export: Yes
Catalogue: Yes
Contact: Shakir Alzaidi
Email: info@green-science.com
Address: Unit 68,
Third Avenue, Deeside Industrial Park,
Deeside, Flintshire CH5 2LA,
Scotland
Tel: 01244 281333
Fax: 01244 281878

Green Shoes
www.greenshoes.co.uk

Specialities: Footwear; vegan
Website: Designers and makers of quality hand made customised footwear in leather as well as special vegan limited edition. Shoes can be customised by choosing the design, the colour, the sole type, the length and the width. There is a resoling service. Website gives details of products including the vegan range. Orders may be placed by post, telephone or email.
Links: Yes
Credit cards: Visa; Mastercard; Access
Email: info@greenshoes.co.uk
Address: 69 High Street, Totnes, Devon TQ9 5PB
Tel: 01803 864 997
Fax: 01803 964 997

The Green Shop
www.greenshop.co.uk

Specialities: Household products; bodycare; paints; energy saving
Website: Environmentally and ethically aware products. Comprehensive coverage of eco-friendly house products, including cleaning materials, toiletries, stationery and paints. Mail order and on-line catalogue for secure ordering. There is a search facility by key words and price range.
Links: Yes
On-line ordering: Yes
Email: mailorder@greenshop.co.uk
Address: Holbrook Garage, Bisley, Stroud, Gloucestershire GL6 7AD
Tel: 01452 770629
Fax: 01452 770629

Green Ways
www.green-ways.co.uk

Specialities: Algae control; gardening
Website: The home of natural algae control. Manufacturers of pond pads using barley straw to control all types of algae including green water and blanket weed in garden ponds. Website explains how they work, gives information on the product range with prices and specifications and provides a listing of suppliers.
Email: sales@green-ways.co.uk
Address: Southend Farm, Long Reach, Ockham, Woking, Surrey GU23 6PF
Tel: 01483 281 391
Fax: 01483 281 392

Green Woodland Burial Services
www.greenburials.co.uk

Specialities: Woodland burial; natural death
Website: Provides an environmentally friendly, cost-effective alternative to traditional burials in graveyards or cemeteries. The website gives details of costs of full internment or internment of ashes in a natural woodland setting and all associated services. There is an email facility for an application form and further details.
Email: gburials@talk21.com
Address: 256 High Street, Harwich, Essex CO12 3PA
Tel: 01255 503 456
Fax: 01255 504 466

Greencuisine Ltd
www.greencuisine.org

Specialities: Cookery courses; nutrition
Website: Website dedicated to organic food and health. Organic nutritional principles evolved from the teaching of chef and nutritionist Daphne Lambert at Penrhos Court (see separate entry). There is a shop selling food, wines and books. There is also information on courses, recipes, articles and a forum. There is a useful page of FAQs with concise factual answers.
Links: Yes
On-line ordering: Yes
Email: info@greencuisine.org
Address: Kington, Herefordshire HR5 3LH
Tel: 01544 230720
Fax: 01544 230754

Greenfibres Organic Textiles
www.greenfibres.com

Specialities: Clothes; fabrics; bedding; household linen; babywear; nappies
Website: Manufactures and sells organic clothes which are both stylish and ethical. Baby clothes, nappies, and babycare products. Supporters of organic agriculture and fair working conditions. Materials include organic cotton, organic linen, organic wool and natural hemp. On-line ordering and mail order are available.
Links: Yes
On-line ordering: Yes
Catalogue: Yes
Contact: William Lana
Email: mail@greenfibres.com
Address: 99 High Street, Totnes, Devon TQ9 5PF
Tel: 01803 868 001
Fax: 01803 868 002

Greenfield Coffins
www.greenfieldcoffins.com

Specialities: Natural death; biodegradable coffins
Website: Maker and supplier of coffins made from 100% biodegradable materials and contact point for natural burial sites. Website gives details and prices of coffins made from recycled cardboard. Special designs are available to order.
Email: mail@greenfieldcoffins.com
Address: 2-6 Lakes Road, Braintree, Essex CM7 3SS
Tel: 01376 327 074
Fax: 01376 342 975

The Greenhouse Edinburgh
www.greenhouse-edinburgh.com

Specialities: Guesthouse; holidays; vegetarian & vegan
Website: Edinburgh's vegan and vegetarian guest house situated in the heart of the historic city. Website gives information on rooms and rates and shows a breakfast menu. There is an email enquiry booking form with tariff. There is also a photographic tour of the guest house available on the website.
Email: niallk@hotmail.com
Address: 14 Hartington Gardens, Edinburgh EH10 4LD, Scotland
Tel: 0131 622 7634

Greenpeace
www.greenpeace.org.uk

Specialities: Organisation; environment; sustainable development
Website: Independent international campaigning organisation combating environmental abuse and promoting positive solutions to destructive practices. Includes action against genetically modified foods. Extensive website gives details of campaigns, how to get involved and information on the media. There is an on-line facility for joining Greenpeace.
Email: info@uk.greenpeace.org
Address: Canonbury Villas, London N1 2PN
Tel: 020 7865 8100
Fax: 020 7865 8200

Greensleeves Clothing
www.greensleevesclothing.com

Speciality: Baby clothes
Website: Stylish organic baby clothes that are kind to babies and toddlers and safe for the environment. Catalogue includes newborn gift boxes, essential organic wool and cotton outfits, socks, tights, cosy organic bodys and nightwear for 0 - 3 years. Website has email request for printed catalogue.
Catalogue: Yes
Contact: Meike Cassens Wadsworth
Email: info@greensleevesclothing.com
Address: 61 Oakwood Road, London NW11 6RJ
Tel: 020 8455 2809
Fax: 020 8455 2809

Greenstuff Ltd
www.greenstuff.org.uk

Speciality: Meat
Website: Irish organic beef, lamb and sausages. The site provides general information and recipes. It is also possible to log on to the website, key in the lot number on a packet of meat and see exactly which farm in Ireland it came from.
Email: info@greenstuff.org.uk
Address: Greenstuff Limited @ Hub TM, Axe & Bottle Court,
70 Newcomen Street, London SE1 1YT
Tel: 020 7378 3443
Fax: 020 7 378 3445

Greenwich Organic Foods
www.greenwichorganic.co.uk

Specialities: Retail outlet; box scheme; groceries
Website: Delivery service available. Very large range of organic foods. On the website the products page gives a full listing of organic produce and groceries, the organics section gives information on organic standards and seasonal availability, and there are delivery details and information on the shop itself. The products page can be used to make an order which is then emailed for confirmation. There is also a recipe section.
Links: Yes
Contact: Catherine Russell
Email: info@greenwichorganic.co.uk
Address: 86 Royal Hill, Greenwich, London SE10 8RT
Tel: 020 8488 6764

Growing Concern Organic Farm
www.growingconcern.co.uk

Specialities: Meat; ready meals
Website: Family run organic farm selling award winning pork pies, organic beef, organic chickens, organic sausages and home farm smoked organic produce. Also organic ready meals, puddings, pies, cakes and bread. Printable order form, telephone ordering and interactive PDF order form. Website also gives background information on the farm and family, their philosophy and a location map.
Contact: Mary & Michael Bell
Email: info@growingconcern.co.uk
Address: Home Farm,
Woodhouse Lane, Nanpantan,
Leicestershire LE11 3YG
Tel: 01509 239228
Fax: 01509 239228

H.F.M.A
www.hfma.co.uk

Specialities: Organisation; health foods; food supplements, herbal remedies; aromatherapy; homoeopathy
Website: The major trade association for the specialist health product industry. Its mission is to build and maintain a successful and credible industry sector and its primary aim is to obtain a good regulatory environment for its member businesses and for the consumer. Website provides information and membership details. The association covers food supplements, health foods, herbal products, natural remedies, homoeopathy, sports nutrition products, natural cosmetics and aromatherapy.
Contact: Penny Viner
Email: pviner@hfma.co.uk
Address:
The Health Food Manufacturers Association,
63 Hampton Court Way,
Thames Ditton, Surrey KT7 0LT
Tel: 020 8398 4066
Fax: 020 8398 5402

Halcyon Seeds
www.halcyonseeds.co.uk

Specialities: Seeds; salad seeds; unusual seeds; gardening
Website: A major UK source for unusual seeds for salad leaves, herbs, organic vegetables and edible flowers. There is also a Grower's Club to give help and information with growing instructions.
Links: Yes
On-line ordering: Yes
Credit cards: Most major cards
Export: Europe
Catalogue: Yes
Contact: Richard Bartlett
Email: enquiries@halcyonseeds.co.uk
Tel: 01865 890180

Hale Clinic
www.haleclinic.com

Speciality: Complementary medicine
Website: The aim is to combine the principles of conventional and complementary medicine as no one system has the whole answer to every medical problem. Over 100 practitioners offering 40 different treatments are based at the clinic, many of whom are multi-disciplinary. The website has sections on ailments, treatments, Hale education, Teresa Hale, breath connection, news and events.
Email: admin@haleclinic.com
Address: 7 Park Crescent, London W1B 1PF
Tel: 020 7631 0156
Fax: 020 7323 1693

Half Moon Healthfoods
www.halfmoon-healthfoods.co.uk

Specialities: Retail shop; box scheme
Website: Wholefood store specialising in organic produce, breads, dairy and wines. Website provides photographs.
Contact: Adrian Midgley
Email: sales@halfmoon-healthfoods.co.uk
Address: 6 Half Moon Street, Huddersfield, West Yorkshire HD1 2JJ
Tel: 01484 456392
Fax: 01484 310161

Hambleden Organic Herb Trading Company
www.hambledenherbs.co.uk

Specialities: Herbs; dried herbs
Website: The only organic dried herb specialists in Britain - over 550 different herbs and spices supplied. The all-organic range includes herbal teas, infusions, culinary herbs and spices, tinctures, Christmas products, carob and henna. Website was under construction at time of writing.
Contact: Gaye Donaldson
Email: info@hambledenherbs.co.uk
Address: Court Farm, Milverton, Somerset TA4 1NF
Tel: 01823 401104
Fax: 01823 401001

Hampshire Farmers' Market
www.hants.gov.uk/farmersmarkets

Speciality: Farmers' markets
Website: Possibly the UK's largest farmers' market held on the last Sunday of each month with up to 8,000 visitors each month. Website gives dates and venues for the current year. It also lists producers and their addresses under product categories.
On-line ordering: No
Address: Middle Brook Street, Winchester, Hampshire
Tel: 01962 845135

The Hampstead Tea and Coffee Company
www.hampsteadtea.com

Specialities: Teas; coffees
Website: Bio-dynamic, organic and fairtrade teas. The company prides itself on the purity and quality of its teas and the natural harmony and well being of each estate and its people. The website provides a history of the company, information on the first bio-dynamic tea estate in the world, and the fairtrade policy. Mail order facilities are available and each product is fully described.
Contact: Kiran Tawadey
Email: info@hampsteadtea.com
Address: PO Box 2448,
London NW11 7DR
Tel: 020 8731 9833
Fax: 020 8458 3947

Harvest Forestry
www.harvestforestry.co.uk

Specialities: Forestry, furniture; groceries; retail shop
Website: Company has a tree surgery, contracting and consultancy, timber and furniture business in addition to a retail shop which is committed to buying locally and also grows its own produce. The shop stocks certified organic fresh fruit, vegetables and herbs. Details of all aspects of the business are shown on the website.
Email:
harvestforestry@fastnet.co.uk
Address: 1 New England Street,
Brighton, East Sussex BN1 4GT
Tel: 01273 689725
Fax: 01273 622727

Hawkshead Organic Trout Farm
www.organicfish.com

Specialities: Trout; fish
Website: The organic trout are fed a diet containing ingredients suitable for organic production and they benefit from an holistic approach. The website provides information on the company and its objectives together with a price list and on-line ordering facility. There is also a newsletter.
On-line ordering: Yes
Credit cards: Most major credit cards
Email: trout@hawkshead.demon.co.uk
Address: The Boathouse,
Hawkshead, Cumbria LA22 0QF
Tel: 01539 436541
Fax: 01539 436541

Halzephron Herb Farm
www.halzherb.com

Specialities: Herbal remedies; herbs
Website: Herbal products free from all additives, preservatives or colourings. Herbal medicine products and herb dips are available by mail order. Many of the food products have unfortunately had to be withdrawn from mail order sale due to their weight cost ratios for postage. The full range of products is available from the shop and location details and opening times are shown on the website.
On-line ordering: Yes
Credit cards: Visa; Mastercard; Switch
Catalogue: Yes
Address: Gunwallowe, Helston,
Cornwall TR12 7QD
Tel: 01326 240 652
Fax: 01326 241 125

Heal Farm Meats
www.healfarm.co.uk

Specialities: Meat; poultry; turkeys
Website: Rare breed meat reared on the basis of natural sustainable farming and a high welfare system of animal production which is both ethically and ecologically sound. Meat processing is done on the farm and delivery nationwide is by a guaranteed overnight service. The website includes a statement of the farm's position on organic farming. There is a product listing, a newsletter, Q&A, and endorsements.
Links: Yes
On-line ordering: Yes
Credit cards: Most major credit cards
Email: enquiries@healfarm.co.uk
Address: King's Nympton, Umberleigh, Devon EX37 9TB
Tel: 01769 574341
Fax: 01769 572839

Healing Herbs
www.healing-herbs.co.uk

Speciality: Herbal remedies
Website: Healing Herb's Bach remedies are the traditional 38 flower remedies discovered by Dr Edward Bach in the 1930s. On the web pages there is background and history, a complete list of the remedies, how to use the remedies and the philosophy behind them. There is also an order form with prices which can be printed out and emailed, posted or faxed.
Credit cards: Visa; Mastercard; Switch; Maestro; JCB
Contact: Julian Barnard
Email: hh@healing-herbs.co.uk
Address: PO Box 65, Hereford, Herefordshire HR2 0DX
Tel: 01873 890218
Fax: 01873 890314

Health Imports
www.health-imports.co.uk

Specialities: Importers; tea tree oil
Website: Website gives information on tea tree oil. There are pages on the many different problems it can help treat, a product listing and on-line ordering, a background to harvesting techniques and ethics, and a UK stockists search service.
Contact: Michael Lightowlers
Email: michael.lightowlers@health-imports.co.uk
Address: Binbrook Mill, Young Street, Bradford, West Yorkshire BD8 9RE
Tel: 01274 488511
Fax: 01274 541121

Health Perception
www.health-perception.co.uk

Speciality: Health supplements
Website: The original company to bring glucosamine to the UK and sponsors of its first UK clinical trial. Website gives a general introduction and then has information and product pages on joint mobility, bone health, brain function, circulation, hormonal balance, the immune system and body and mind. Products may be ordered on-line
On-line ordering: Yes
Contact: Helen Issacs
Email: sales@health-perception.co.uk
Address: Unit 12, Lakeside Business Park, Swan Lane, Sandhurst, Berkshire GU47 7DN
Tel: 01252 861454
Fax: 01252 861455

The Health Store
www.thehealthstore.co.uk

Specialities: Wholesaler; buying group
Website: Leading buying group and wholesaler for independent health food retailers. Over 6,500 chilled frozen and ambient lines supplied. Web pages gives details of history and ethos, the retail support service and support team, and examples of the product ranges. There is information on how to become a shareholding member.
Contact: Graeme Gunn
Email: mail@thehealthstore.co.uk
Address: Bastow House, Queens Road, Nottingham, Nottinghamshire NG2 3AS
Tel: 0115 955 5255
Fax: 0115 955 5290

Healthaid
www.healthaid.co.uk

Speciality: Food supplements
Website: A leading vitamin and supplement brand in the UK. Web pages provide a company profile, product range listing, prices, nutrient guide and new product listing.
Contact: Sanjit Patel
Email: sales@pharmadass.com
Address: Pharmadass Ltd, 16 Aintree Road, Greenford, Middlesex UB6 7LA
Tel: 020 8991 0035
Fax: 020 8997 3490

Healthquest
www.healthquest.co.uk

Specialities: Bodycare; nutrition
Website: Healthcare has created a catalogue of products focused on staying well, fit and relaxed. There is an email catalogue request form. Items are available for on-line ordering.
On-line ordering: Yes
Catalogue: Yes
Contact: Nemish Mehta
Email: info@healthquest.co.uk
Address: Healthquest, 7 Brampton Road, Kingsbury, London NW9 0BX
Tel: 020 8206 2066
Fax: 020 8206 2022

Healthquest - Organic Blue
www.organicblue.com

Specialities: Body care; cosmetics
Website: Organic ingredients chosen from traditional medicine systems blended into superb bases for the natural health market. Sophisticated looking website with pages on philosophy and product range. Contact details provided.
Contact: Nemish Mehta
Email: info@healthquest.co.uk
Address: 7 Brampton Road, Kingsbury, London NW9 9BX
Tel: 020 8206 2066
Fax: 020 8206 2022

Healthy Flooring Network
www.healthyflooring.org

Specialities: Allergies; asthma; flooring
Website: An alliance of individuals and organisations concerned about health, asthma and allergies. It aims to raise awareness of the links between fitted carpets, PVC flooring and health and to encourage and promote alternatives. Website provides a statement of concern, information, reports, latest news and press releases. There is a guide to alternative flooring suppliers with full contact details.
Links: Yes
Email: info@healthyflooring.org
Address: PO Box 30626, London E1 1TZ
Tel: 020 7481 9004
Fax: 020 7481 9144

The Healthy House
www.healthy-house.co.uk

Specialities: Allergies; babycare; bodycare; household; paints; clothing
Website: Specialises in products for people who suffer from allergies and asthma, and includes organic cotton bedding, air purifiers and allergy-free carpets. Website gives details and descriptions of relevant allergies. There is an on-line shopping facility with a comprehensive index of product categories covering babies, bodycare, household, clothing, bedding, paints, and books.
Links: Yes
On-line ordering: Yes
Catalogue: Yes
Contact: Maxima & Don Skelton
Email: info@healthy-house.co.uk
Address: Cold Harbour, Ruscombe, Stroud, Gloucestershire GL6 4DA
Tel: 01453 752 216
Fax: 01453 753 533

Hemp Food Industries Association
www.hemp.co.uk

Speciality: Hemp foods
Website: Independent hemp community focusing on hemp-based food products. Website gives details of products, the benefits of hemp, newsroom, fun stuff, HFIA, and links.
Links: Yes
Contact: Paul Benham
Email: hfia@hemp.co.uk
Address: PO Box 204, Barnet, Hertfordshire EN4 8ZQ
Tel: 07050 600418
Fax: 07050 600419

Hemp Union
www.hemp-union.karoo.net

Specialities: Hemp products; clothing; cosmetics; hemp foods
Website: Leading UK hemp company for both consumer and industrial products. Website provides information, links and shopping relating to hemp. There is a full illustrated product listing but at the time of writing no e-commerce facility. Orders can be placed by phone, fax or post. The website includes publication lists and on-line recipes.
Credit cards: Most major credit cards
Email: sales@hemp-union.karoo.co.uk
Address: 24 Anlaby Road, Hull, East Yorkshire HU1 2PA
Tel: 01482 225 328

Henry Doubleday Research Association
www.hdra.org.uk

Specialities: Organisation; gardening; agriculture; recycling; composting; heritage seed library
Website: Europe's largest organic membership organisation. It provides information, advice on organic subjects; it offers consultancy; it runs demonstration gardens; it has a restaurant and a shop; it runs a vegetable box scheme. The association researches organic gardening, farming and food and acts as a consultancy for recycling, composting, organic garden design, landscaping, and organic retailing and catering. The extensive website provides information pages on organic gardening, guides and fact sheets and details of courses amongst a wealth of other useful pages. The HDRA Organic Food Club, Wine Club and gardening catalogue can all be accessed for on-line ordering.
On-line ordering: Yes
Contact: Jackie Gear
Email: enquiry@hdra.org.uk
Address: Ryton Organic Gardens, Ryton on Dunsmore, Coventry, Warwickshire CV8 3LG
Tel: 02476 303517
Fax: 02476 639229

Herb Society
www.herbsociety.co.uk

Specialities: Herbs; organisation
Website: The Society aims to increase the understanding, use and appreciation of herbs and their benefits to health. It is an educational charity disseminating information about herbs. The website provides a number of articles, information on the *Herbs* magazine and its contents, membership information and events listing.
Links: Yes
Email: email@herbsociety.co.uk
Address: Deddington Hill Farm, Warmington, Banbury, Oxfordshire OX17 1XB
Tel: 01295 692 000
Fax: 01295 692 004

Herbaticus
www.herbaticus.co.uk

Specialities: Animal care; food supplements for animals
Website: Natural herbal food supplements for horses, dogs and cats. There is a listing of products by animal and health condition with full details of treatment and dosage on the website. Orders can be placed by phone, fax, post or email. There is a trade price list for vets.
Links: Yes
Credit cards: All major credit cards
Catalogue: Free literature
Email: enquiries@herbaticus.co.uk
Address: 76 Snape Hill Lane, Dronfield, Derbyshire S18 2GP
Tel: 01246 411 949
Fax: 01246 418 482

Herbon
www.herbon.com

Speciality: Herbal remedies
Website: Herbon echinacea-based herbal lozenges offer symptom-specific relief whilst boosting the immune system. The website is that of the parent company in Canada. It gives details of the products, the company and research but availability information is not UK oriented.
Contact: Christina Tattersall
Address: Binbrook Mill, Young Street, Bradford, West Yorkshire BD8 9RE
Tel: 01274 488511
Fax: 01274 541121

Herbs Hand Healing
www.herbshandshealing.co.uk

Specialities: Herbs; herbal remedies
Website: A mail order company and a herbal information service, dispensary and clinic. High quality organic herbs, correct dosages, high standards of manufacturing and well-researched modern formulae. Products available for sale on-line include herbal formulae, herbal teas, colon products, superfood, powders and pessaries, cleansers, skincare, haircare, oils and home cleaners. Website also provides articles and news items. Information on contraindications and drug interactions is available on demand.
Links: Yes
On-line ordering: Yes
Catalogue: Herb catalogue
Email: herbsinfo@aol.com
Address: Station Warehouse, Station Road, Pulham Market, Norfolk IP21 4XF
Tel: 0845 345 3727
Fax: 01379 608 201

Higher Hacknell Farm
www.higherhacknell.co.uk

Speciality: Meat
Website: Organic beef and lamb. Nationwide delivery available direct from the farm to the home. The website provides a price list and order form. There is a newsletter and also a recipe page.
Credit cards: Most major cards
Contact: Jo Budden
Email: enquiries@hacknell.fsbusiness.co.uk
Address: Burrington, Umberleigh, Devon EX37 9LX
Tel: 01769 560909

Higher Riscombe Farm
www.higherriscombefarm.co.uk

Specialities: Bed & breakfast; holidays
Website: Bed and breakfast on organic farm - organic breakfasts and evening meals available. Website provides information on the farm and activities available, the weather and a newsletter.
Links: Yes
Contact: Rona Ayres
Email: intray@higherriscombefarm.co.uk
Address: Exford, Nr Minehead, Somerset TA24 7JY
Tel: 01643 831184
Fax: 01643 831184

Highland Organic Foods
www.highlandorganics.co.uk

Speciality: Meat
Website: Producers and distributors of organic meat from Wales for over 15 years. The website gives background information on the farm and lists the organic products with prices. There is delivery throughout the UK.
Credit cards: Visa; Mastercard
Catalogue: Price list available
Email: sales@highlandorganics.co.uk
Address: 4 Bittacy Hill, Mill Hill, London NW7 1LB
Tel: 020 8371 0770
Fax: 020 8349 4623

Hill House Retreats: Centre for Wholistic Living
www.hillhouseretreats.co.uk

Specialities: Retreats; vegetarian & vegan
Website: It is intended to be a place that helps to dissolve the illusion of separation between inner spiritual life and outer worldly life where we can renew our connection to ourselves, to nature and to life itself. The website provides pictures of the house and rooms as well as the locality. Individual retreats, workshops and celebrations, therapies and outdoor activities are available. There is a page showing the current programme.
Email: infoweb@hillhouseretreats.co..uk
Address: Greenwood Courses, The Church Hall, Hill House, Llanstefan, Carmarthen SA33 5JG, Wales
Tel: 01267 241 999
Fax: 01267 241 999

Hilton Herbs
www.hiltonherbs.com

Speciality: Animal care
Website: Homoeopathy for horses, comfrey for canines, herbal remedies for animals. The website gives a history of the company, a helpline, quality control, information on ethical and environmental issues and news pages. There is on-line ordering with a product search facility,
On-line ordering: Yes
Credit cards: Most major credit cards
Export: Yes
Contact: Hilary & Tony Self
Email: helpline@hiltonherbs.com
Address: Downclose Farm, North Perrott, Crewkerne, Somerset TA18 7SH
Tel: 01460 78300
Fax: 01460 78302

Hi Peak Feeds (Proctors) Bakewell Ltd
www.hipeak.co.uk

Specialities: Animal feeds; dog food
Website: Organic non-GMO animal feeds manufactured and supplied to farmers and smallholders for farm livestock, horses and dogs. Nationwide delivery and mail order service. A subsidiary company Hi Peak Organic Food meets the consumer demand for organic food with poultry, meat and eggs. Website provides general company and product information. It also has information for potential organic suppliers and producers.

Contact: John Walker
Email: info@hipeak.co.uk
Address: Hi Peak Feeds Mill, Sheffield Road, Killamarsh, Derbyshire DE22 3AA
Tel: 0114 248 0608
Fax: 0114 247 5189

Hipp Organic
www.hipp.co.uk

Specialities: Baby food; maternity
Website: Organic baby food and formulas. The website provides a history of the company and its organic commitment, a product listing with age level suitability and a page of FAQs on breastfeeding. There is a selection facility for special nutrition requirements related to allergies and incompatibilities. There is also a page of useful addresses of relevant associations and organisations.
Address: 165 Main Street, New Greenham Park, Newbury, Berkshire RG19 6HN
Tel: 01635 528250
Fax: 01635 528271

Hippychick
www.hippychickltd.co.uk

Specialities: Childcare; babycare; baby shoes; baby carriers
Website: Simple, practical and healthy products to help care for children including the Child Hip Seat and Armakids range of natural products for children which include massage oils, bubble baths, soaps and creams. The range is available through retail outlets and may also be purchased on-line through the website or through the printable order form.
On-line ordering: Yes
Credit cards: Visa; Mastercard; Switch; Delta
Contact: Julia Minchin
Email: info@hippychickltd.co.uk
Address: Barford Gables, Spaxton, Bridgwater, Somerset TA5 1AE
Tel: 01278 671 461
Fax: 01278 671 715

Hockerton Housing Project
www.hockerton.demon.co.uk

Specialities: Ecological housing; housing
Website: The UK's first earth-sheltered, self sufficient ecological housing development that works. The families in the five houses generate their own energy, harvest their water and recycle waste without any pollution or carbon monoxide emissions. Tours take place once a month. The website provides information on the project, media information, publications and services which can be ordered on a printable order form, and a virtual tour of the project.
Links: Yes
Email: hhp@hockerton.demon.co.uk
Address: The Watershed, Gables Drive, Hockerton, Southwell, Nottinghamshire NG24 0PQ
Tel: 01636 816902

Holden Meehan IFA Ltd
www.holden-meehan.co.uk

Specialities: Ethical investment; financial
Website: Independent advisers specialising in green and ethical finance planning. Website provides a helpful introductory page. There is also a green questionnaire which will help working out whether the services offered meet the customer's requirements. The free *Guide to Socially Responsible Forms of Investment* can be ordered on-line.
Catalogue: Free independent guide
Email: info-web@holden-meehan.co.uk
Address: New Penderel House, 283-287 High Holborn, London WC1V 7HP
Tel: 020 7692 1700
Fax: 020 7692 1701

Holgran
www.holgran.co.uk

Specialities: Bakery ingredients; food industry supplier
Website: UK manufacturer of organic malted ingredients, concentrates and mixes for the production of premium bread products. Website gives details of the mixes and the customer service available.
Contact: Alan Marson
Email: bits4bread@holgran.co.uk
Address: Granary House, Wetmore Road, Burton on Trent, Staffordshire DE14 1TE
Tel: 01283 511255
Fax: 01283 511220

Home Education Advisory Service (HEAS)
www.heas.org.uk

Specialities: Home education; organisation
Website: An opportunity to explore new educational possibilities offering diversity, creativity and an individual approach to learning. The service is committed to providing information about home education to everyone. Website covers what it offers families, what it offers LEAs, and its publications and services. There are links to useful websites.
Links: Yes
Email: admin@heas.org.uk
Address: PO Box 98, Welwyn Garden City, Hertfordshire AL8 6AN
Tel: 01707 371 854
Fax: 01707 371 854

Home Place Farm
www.holidayexmoor.co.uk

Specialities: Bed & breakfast; holiday cottages; holidays
Website: The holiday business is run in an environmentally friendly way. Website gives details of the holiday cottages with photos and information on bed and breakfast, the private spa, the local surroundings, activities, facilities, prices and booking information.
Contact: Mark & Sarah Ravenscroft
Email: markandsarah@holidayexmoor.co.uk
Address: Challacombe, Barnstaple, Devon EX31 4TS
Tel: 01598 763 283

Home Farm Deliveries
www.homefarm.co.uk

Specialities: Groceries; home delivery
Website: A wide range of all categories of organic food delivered direct to homes in Manchester and surrounding counties, acting as a link between farmers and growers and consumers. A brochure can be ordered which details the products available but there is also a general category description on the website.
On-line ordering: Yes
Catalogue: Brochure
Contact: Mike Shaw
Email: info@homefarm.co.uk
Address: Studio 19, Imex Business Park, Hamilton Road, Manchester M13 0PD
Tel: 0161 224 8884
Fax: 0161 224 8826

Honesty Cosmetics
www.honestycosmetics.co.uk

Specialities: Skincare; haircare; bodycare; vegetarian & vegan
Website: Mail order catalogue of skin and hair care products all of which are suitable for vegetarian and vegans. The site gives product listings and on-line ordering was under development at time of writing. Orders can sent or faxed on the printable order form with prices.
Links: Yes
Credit cards: All major credit cards
Email: info@honestycosmetics.co.uk
Address: Lumford Mill, Bakewell, Derbyshire DE45 1GS
Tel: 01629 814888
Fax: 01629 814111

Honeyrose Products Ltd
www.honeyrose.com

Specialities: Herbal remedies; tobacco alternatives; skincare; babycare
Website: Nicotine-free, tobacco-free, non-addictive alternative herbal cigarettes and smoking mixes as an aid to giving up smoking. Made from pure natural plants flavoured with honey and fruit juices. Website provides stop-smoking information and details of the other products available. There is a link for on-line buying with credit cards and there is a printable order form.
Contact: John Heywood
Email: honeyrose.products@virgin.net
Address: Creeting Road, Stowmarket, Suffolk IP14 5AY
Tel: 01449 612137
Fax: 01449 770409

Horizon Organic Dairy
www.horizonorganic.co.uk

Speciality: Dairy
Website: Dedicated organic dairy which uses no artificial pesticides or fertilisers on the land and bans GMOs. Range of organic skimmed, semi-skimmed, wholemilk, and non-homogenised milk. A product listing is given on the website.
Links: Yes
Contact: David Stacey
Email: d.stacey@dial.pipex.com
Address: Leigh Manor, Minsterley, Shropshire SY5 0EX
Tel: 01743 891990
Fax: 01743 891929

Horticultural Correspondence College
www.hccollege.co.uk

Specialities: Courses; gardening; herbs; conservation
Website: Home study courses including organic gardening, conservation studies, garden design, leisure gardening and herbs for pleasure and profit. Also prepares for RHS exams and Royal Forestry Society exams. The website provides information on the courses and enrolment and provides an application form. There is a page for FAQs and a general introduction to the college.
Email: info@hccollege.co.uk
Address: Little Notton Farmhouse, 16 Notton, Lacock, Chippenham, Wiltshire SN15 2NF
Tel: 01249 730326
Fax: 01249 730326

Hotel Mocking Bird Hill
www.hotelmockingbirdhill.com

Specialities: Hotel; holidays
Website: Hotel with eco-friendly philosophy which provides tips, background information and insight into Jamaican arts and culture. Website gives illustrated details of the accommodation, the restaurant which uses only fresh local produce including vegetarian selections, activities, location and rates.
Email: mockbird@cwjamaica.com
Address: PO Box 254, Port Antonio, Jamaica
Tel: + 876 993 7267
Fax: + 876 993 7133

Howbarrow Organic Farm
www.howbarroworganic .demon.co.uk

Specialities: Meat; poultry; vegetables; soft fruit; box scheme; bed & breakfast
Website: Produce produced on smallholding without herbicides, pesticides, chemical fertilisers, antibiotics, growth promoters or GMOs. It supplies local market stall, farm shop and box scheme. Bed and breakfast is available. The site provides general information, photographs, news and a produce list. A fax form is available on-line for printing.
Links: Yes
Contact: Julia Sayburn
Email: enquiries@howbarroworganic.demon.co.uk
Address: Cartmel, Grange over Sands, Cumbria LA11 7SS
Tel: 015395 36330
Fax: 015395 36330

Huggababy Natural Baby Products
www.huggababy.co.uk

Specialities: Baby carriers; babycare; wooden toys; maternity
Website: Huggababy own-brand products are as natural as possible and products from other companies are chosen with great care. Products range from a baby carrier to strollers, cradle cushions, wooden toys, natural lambskins and a maternity bra. Items can be ordered through a secure server or by phone, fax or post. All prices include UK p&p.
On-line ordering: Yes
Credit cards: Visa; Mastercard; Switch
Contact: Margaret Hastings
Email: margaret@huggababy.co.uk
Address: Great House Barn, New Street, Talgarth, Brecon, Powys LD3 0AH, Wales
Tel: 01874 711 629
Fax: 01874 711 037

Hugger-Mugger Yoga Products
www.yoga.co.uk

Specialities: Yoga equipment; yoga; holidays
Website: Website offers on-line ordering for a complete range of mats, blocks, belts, clothing, bolsters, books and videos. This is also the site for the Yoga Centre and there is information on yoga holidays in Crete, Thailand, Turkey, Sri Lanka and Scotland as well as details of yoga courses.
Email: yoganet@ednet.co.uk
Address: 12 Roseneath Place, Edinburgh EH9 1JB, Scotland
Tel: 0131 221 9977
Fax: 0131 221 9697

Humberside Nursery Products Co
www.hnpdirect.com

Specialities: Seeds; fertilisers; composts; gardening
Website: Product range is listed on the website under these headings: substrate, nutrition, hygiene, seeds and sundries. An on-line email order form may be used to get quotations on orders. There is additional advice on organic pest control on the website.
Links: Yes
Credit cards: Most major credit cards
Email: sales@hnpdirect.com
Address: Common Lane, North Cave, East Yorkshire HU16 2PE
Tel: 01430 423065

ICOF Ltd
www.icof.co.uk

Specialities: Financial advice; loan finance
Website: Industrial Common Ownership Finance provides loan finance for co-operatives, employee owned businesses and social enterprises. Website provides information on ethical policy, the ICOF story, and ICOF news. It also gives details of services and borrowing. There are offices in Wales and Cambridge. There is also information on its sister company the Industrial Common Ownership Movement Ltd.
Links: Yes
Email: icof@icof.co.uk
Address: 227C City Road, London EC1V 1JT
Tel: 020 7251 6181
Fax: 020 7336 7407

Indian Champissage Head Massage
www.indianchampissage.com

Specialities: Indian head massage
Website: Website explains the principles of Indian head massage and provides information on services at the centre. Details are given of training courses in champissage and facial rejuvenation. A book is available. There is a list of qualified therapists in the UK with contact details.
Contact: Narendra Mehta
Email: mehta@indianchampissage.com
Address: 136 Holloway Road, London N7 8DD
Tel: 020 7609 3590
Fax: 020 7607 4228

Indigo Herbal Ltd
www.indigoherbal.co.uk

Specialities: Ayurveda; herbal remedies; courses; books
Website: Ayurveda-Go Herbal range. Also tailor-made Vedic courses organised for all backgrounds. Website being re-vamped at time of writing. The current site has a product listing, priced order form for printing and mailing, faxing or emailing. On-line ordering is being introduced. There is a general introductory page on ayurveda.
Contact: Nadine Sayer
Email: info@indigoherbal.co.uk
Address: PO Box 22317,
London W13 8PG
Tel: 020 8621 3633
Fax: 020 8621 3633

Infinity Foods
www.infinityfoods.co.uk

Specialities: Wholesaler; retail shop; bakery; café
Website: Leading national organic wholesaler with over 2,000 organic lines including major brands. It is run as a worker's co-operative. The retail shop carries over 1,500 organic lines. Website covers the wholesale and retail business, the café, recipes and a general products summary. There is a page of terms and conditions.
Catalogue: Yes
Contact: Charlie Booth**Email:** info@infinityfoods.co.uk
Address: Infinity Foods Co-operative,
67 Norway Street, Portslade,
East Sussex BN41 1AE
Tel: 01273 424060
Fax: 01273 417739

Innocent Ltd
www.innocentdrinks.co.uk

Specialities: Smoothies; thickies
Website: Smoothies made from 100% pure crushed fruit with no concentrated juices or additives. Thickies are made from fresh yogurt, fruit, honey and absolutely nothing else. Website gives information on the history of the company and the drinks. There is a listing of stockists and on-line ordering facility.
On-line ordering: Yes
Email: hello@innocentdrinks.co.uk
Address: Fruit Towers,
The Goldhawk Estate,
Brackenbury Road,
London W6 0BA
Tel: 020 8600 3939
Fax: 020 8600 3940

Institute of Traditional Herbal Medicine and Aromatherapy
www.aromatherapy-studies.com

Specialities: Aromatherapy; courses; practitioner training
Website: Website provides information on the institute and on the courses available which include a diploma course in London and a residential diploma course in France. There are also courses on aromatic acupressure, first aid and natural remedies, natural nutrition and Indian head massage. There is a register of qualified aromatherapists. Essential organic oils may be purchased and a full catalogue is available.
Links: Yes
Contact: Gabriel Mojay
Email: info@aromatherapy-studies.com
Address: 35 California Road, Mistley, Essex CO11 1JA
Tel: 01206 393465

INTECAM LTD
www.intecam.com

Specialities: Retail information system; complementary medicine
Website: Camios is an integrated complementary alternative medicine ordering and information system for pharmacies and healthstores. It provides instant access to information. Website provides general information about the system and installation and further details can be requested on an email form.
Contact: Heather Sutton
Email: intecam.com@virgin.net
Address: Mena, 29 Steyne Road, Bembridge, Isle of Wight PO35 5UL
Tel: 01983 875527
Fax: 01983 875527

INTERNATIONAL BEE RESEARCH ASSOCIATION
www.cf.ac.uk/ibra

Specialities: Bees; apiculture
Website: Association exists to promote the worldwide study and conservation of bees which in themselves are indicators of the world's biodiversity. Publishers of *Honey and Healing*. Website explains how to be part of IBRA, gives information on publications, the staff, conferences, the council and an international bee people directory. There is a membership application form on-line.
Links: Yes
Email: ibra@cardiff.ac.uk
Address: 18 North Road, Cardiff CF10 3DT, Wales
Tel: 02920 372409
Fax: 02920 665522

INTERNATIONAL FEDERATION OF AROMATHERAPISTS (IFA)
www.int-fed-aromatherapy.co.uk

Specialities: Organisation; aromatherapy
Website: Website giving comprehensive information about the federation and about the practice of professional aromatherapy. Site contains membership information, a directory of members searchable by keywords, a listing of schools with contact details, and pages on research and publications.
Email: i.f.a@ic24.net
Address: Stamford House, 184 Chiswick High Road, London W4 1TH
Tel: 020 8742 2605
Fax: 020 8742 2606

INTERNATIONAL FEDERATION OF REFLEXOLOGISTS
www.reflexology-ifr.com

Specialities: Reflexology; organisation
Website: Website gives an introduction to the history of reflexology in the UK and details on conferences, workshops and events. There is an FAQ page.
Links: Yes
Contact: Renee Tanner
Email: ifr44@aol.com
Address: 76-78 Edridge Road, Croydon, Surrey CR0 1EF
Tel: 020 8667 9458
Fax: 020 8649 9291

INTERNATIONAL INSTITUTE OF REFLEXOLOGY (UK)
www.reflexology-uk.co.uk

Specialities: Reflexology; courses; practitioner training
Website: Website provides a brief definition and history of reflexology and an introduction to the institute. There are details of training courses held in the UK, accredited by the Open and Distance Learning Council. There is a practitioner register of IIR parishioners on a regional basis. A list of books and charts available for purchase from the institute is included.
Links: Yes
Email: reflexology_uk@hotmail.com
Address: 255 Turleigh, Bradford on Avon, Wiltshire BA15 2HG
Tel: 01225 865899
Fax: 01225 865899

Irish Organic Farmers & Growers Association
www.irishorganic.ie

Specialities: Organisation; certification; farming
Website: IOFGA, the Irish certifying body, is open to farmers, growers, consumers and others interested in the production of healthy food and the protection of the environment. It publishes a magazine, *Organic Matters*. Website under construction.
Contact: Eva Draper
Email: gibneyiofga@eircom.net
Address: Organic Farm Centre, Harbour Road, Kilbeggan, West Meath, Ireland
Tel: + 353 (0) 506 32563
Fax: + 353 (0) 506 32063

Irma Fingal-Rock
www.rockwines.co.uk

Specialities: Cheese; wines; olive oils
Website: Importers and merchants of fine wines - an eclectic selection from around the world - but specialise in Burgundy wines. Mail order with nationwide delivery. On-line catalogue of wines, including organic, with prices. Ordering information page.
Contact: Tom Innes
Email: rockwines@lineone.net
Address: 64 Monnow Street, Monmouth, Gwent NP5 3EN, Wales
Tel: 01600 712372
Fax: 01600 712372

Jac by the Stowl - Humungus Fungus
www.jac-by-the-stowl.co.uk

Specialities: Shitake mushrooms; mushroom spawn
Website: Organic shitake mushrooms and organic mushroom spawn. Website gives extensive information on logs, myco-forestry, mycological landscaping, patches, medicinal mushrooms, myco-art, business opportunities and conservation.
Catalogue: Brochure
Contact: Edwards family
Email: enquiries@jac-by-the-stowl.co.uk
Address: Pehrhiw House, Llanddeusant, Llangandog, Carmarthenshire SA19 9YW, Wales
Tel: 01550 740306
Fax: 01550 740306

James White
www.jameswhite.co.uk

Specialities: Juices; apple juice
Website: Freshly pressed organic juices including raspberry, pear, cranberry and apple, carrot and apple, pear and apple. Website provides an on-line shop for fruit juices, fruit coulis, apple juice, organic juices, and eccentrics (spicy tomato juice, spiced ginger aperitif etc.)
On-line ordering: Yes
Credit cards: Most major credit cards
Contact: Lawrence Mallinson
Email: info@jameswhite.co.uk
Address: Whites Fruit Farm, Ashbocking, Ipswich, Suffolk IP6 9JS
Tel: 01473 890111
Fax: 01473 890001

Jekka's Herb Farm
www.jekkasherbfarm.com

Specialities: Herbs; medicinal herbs; culinary herbs; wildflowers; seeds; gardening
Website: Organic herb farm licensed by the Soil Association to produce transplants suitable for organic growing systems. Full catalogue may be downloaded from the website. There is a botanical plant list with Latin and common names. Account holders may order on-line by email.
Links: No
Credit cards: Visa; Mastercard
Export: Europe - individual quotations
Catalogue: Yes (2 - 1st class stamps)
Contact: Jekka McVicar
Email: farm@jekkasherbfarm.com
Address: Rose Cottage, Shellards Lane, Bristol B35 3SY
Tel: 01454 418878
Fax: 01454 411988

Jordans
www.jordanscereals.co.uk

Specialities: Cereals; snacks
Website: UK based, family owned company which manufactures healthy, natural foods including organic breakfast cereals, cereal bars and snack foods. Website gives information on the company, a listing of products with descriptions, and information. There is also information on diet, healthy lifestyles, exercise, and getting active.
Contact: Charles Formby
Address: W Jordan (Cereals) Ltd, Holme Mills, Biggleswade, Bedfordshire SG18 9JY
Tel: 01767 318222
Fax: 01767 600695

Jurlique
www.jurlique.com

Specialities: Skincare; babycare
Website: Naturopathic health and beauty products. The international headquarters of the company are in South Australia. Website gives company history and profile. The on-line store and product catalogue on the website is for US customers.
Email: info@jurlique.com.au
Address:
Naturopathic Health & Beauty Company, Willowtree Marina, West Quay Drive, Yeading, Middlesex UB4 9TB
Tel: 020 8841 6644
Fax: 020 8841 7557

Just Pure
www.justpure.de

Speciality: Skincare
Website: Website is in English and German, dealing with fitness, beauty and nutrition. There is an extensive page on the philosophy of the company and the influence of the moon. A moon calendar features on the site. The product listing gives prices in Deutsch marks.
On-line ordering: Yes
Email: gaby-just@t-online.de
Address: c/o Browns, 23 - 27 South Molton Street, London W1Y 1DA
Tel: 020 7491 7833

Just Wholefoods
www.justwholefoods.co.uk

Specialities: Healthfoods; vegetarian
Website: Vegetarian, organic and healthfoods. Soups, burgers, meals, organic basics and jellies all feature on the website. There is also information on dietary data.
Contact: Anne Madden
Email: james@justwholefoods.co.uk
Address: Unit 16, Cirencester Business Estate, Elliott Road, Love Lane, Cirencester, Gloucestershire GL7 1YS
Tel: 01285 651910
Fax: 01285 650266

Kelly Turkey Farms Ltd
www.kelly-turkeys.com

Specialities: Turkeys; poultry
Website: Traditional natural methods on family run farm to produce top quality turkeys especially for the Christmas market. They re-introduced the bronze feathered turkey. There is an on-line order form. Recipes and tips are provided on-line.
Credit cards: Visa; Mastercard; Access; Eurocard
Email: info@kelly-turkeys.com
Address: Springate Farms, Bicknacre Road, Danbury, Essex CM3 4EP
Tel: 01245 223581
Fax: 01245 226124

Kettle Organics & Kettle Chips
www.kettlefoods.co.uk

Specialities: Snacks; potato chips
Website: Kettle Foods Ltd was founded in Oregon in 1978 with the aim of producing natural, hand-cooked quality foods. Production started in the UK in 1989. Organic, gourmet potato chips are now available nationwide through retailers. Website offers opportunity to join up and keep informed on offers, competitions and recipes. There is a page on organics. A chart showing which supermarkets stock which products is given.
Contact: Claire Last
Email: info@kettlefoods.co.uk
Address: Kettle Foods Ltd, Barnard Road, Bowthorpe, Norwich, Norfolk NR5 9JB
Tel: 01603 744788
Fax: 01603 740375

Ki Kai Shiatsu Centre
www.kikai.freeserve.co.uk

Specialities: Shiatsu; courses; practitioner training
Website: The Centre offers shiatsu treatments from experienced practitioners. It also has a full range of training courses from introductory to full professional qualifications and post-graduate studies. Details of the courses and teachers are available on the website.
Email: love.light@virgin.net
Address: 172a Arlington Road, Camden, London NW1 7HL
Tel: 020 8363 9050

Kids in Comfort
www.slingeasy.co.uk

Speciality: Baby carriers
Website: A baby sling that is both safe and comfortable for parent and baby. Website shows the different fabrics available and the different ways in which the sling can be worn. There is an on-line order form.
On-line ordering: Yes
Credit cards: Switch; Delta
Contact: Karen Lloyd
Email: karen@slingeasy.co.uk
Address: 172 Victoria Road, Wargrave, Reading, Berkshire RG10 8AJ
Tel: 0118 940 4942
Fax: 0118 940 4942

Kids Organic Club
www.kids.organic.org

Specialities: Children's club; education
Website: Competitions, information and games. There is a fact site on the website with information on organic food, farming, soil matters, animal health and welfare, and nutritional content chart. There are also pages for both parents and teachers. There is a feedback form for comments.
Links: Yes
Email: clubinfo@pure.organics.org
Address: Pure Organics Ltd., Stockport Farm, Stockport Road, Amesbury, Wiltshire SP4 7LN
Tel: 01980 626263
Fax: 01980 626264

Kidz Organic Kitchen
www.kidzorganickitchen.com

Speciality: Ready meals
Website: Organic ready meals for children aged 1 - 8. Website was under construction at time of writing.
Contact: Jane Green Armytage
Email: info@kidzorganickitchen.com
Address: Sunshine Organics Ltd, PO Box 2420, Southam, Warwickshire CV47 9YA
Tel: 01926 633086
Fax: 01926 633639

Kindred Spirit
www.kindredspirit.co.uk

Speciality: Magazine
Website: The leading UK Mind, Body and Spirit magazine. It also has a large readership in the US. Website has pages on health and healing and personal development. There is information on back issues and there are plans to make out-of-print articles available on-line to members.
Contact: Nigel Moore
Email: mail@kindredspirit.co.uk
Address: Foxhole, Dartington, Totnes, Devon TQ9 6ED
Tel: 01803 866686
Fax: 01803 866591

Kinesis Nutraceuticals Ltd
www.immunecare.co.uk

Specialities: Food supplements; vitamins; skin care; sports supplements
Website: In addition to physically boosting the body's immune system, the Immunecare range aids fat burning and muscle building, as well as accelerating the injury healing process, and promoting higher levels of stamina and vitality. Products are listed and described on the web pages. Stockist enquiries may be sent by email. There is a page of testimonials.
Contact: Neil Wootten
Email: admin@immunecare.co.uk
Address: 15 Victoria Road, Saltash, Cornwall PL12 4DL
Tel: 01752 203018
Fax: 01752 203018

Kingfisher Natural Toothpaste
www.kingfishertoothpaste.com

Specialities: Toothpaste; vegetarian & vegan
Website: Kingfisher Natural Toothpaste is produced without artificial sweeteners, flavourings, colouring or preservatives. It is vegan and cruelty-free. It is available with or without fluoride. There is a history of Kingfisher on the web pages together with a list of products, prices and stockist information. Technical specifications are given. New products are described.
Contact: Richard Austin
Email: sales@kingfishertoothpaste.com
Address: 21 White Lodge Estate, Hall Road, Norwich, Norfolk NR4 6DG
Tel: 01603 630484
Fax: 01603 664066

Kinvara Smoked Salmon Ltd
www.kinvara-smoked-salmon.com

Speciality: Smoked salmon
Website: Organic salmon reared in the crystal clear waters of the Atlantic Ocean. Their feed is derived from the by-products of herring and mackerel. There is a strong tidal exchange preventing the accumulation of waste and parasites and stocking densities are half that of conventional farms. Salmon fillets are smoked and cured in the traditional way. Smoked salmon, trout and mackerel can be ordered on-line and there is information on organic and wild salmon and the smoking process.
Links: Yes
On-line ordering: Yes
Credit cards: Most major cards
Contact: Declan Droney
Email: smokedsalmon@eircom.net
Address: Kinvara, Co Galway, Ireland
Tel: +353 916 37489
Fax: + 353 916 38193

Kiss My Face
www.kissmyface.com

Specialities: Body care; skincare; suncare; haircare
Website: Kiss My Face natural and organic products are for the whole body. None of the products is animal tested. On-line shopping is for US customers only. There is a page on the site relevant to UK customers. Kiss Organic products are covered in the on-line catalogue.
Links: Yes
Contact: David Davis
Email: info@milfordcollection.com
Address: The Milford Collection, Unit 12, Vastre Estate, Newtown, Powys SY16 1DZ, Wales
Tel: 01686 629919
Fax: 01686 623918

Koppert UK Ltd
www.koppert.nl

Speciality: Pest controls
Website: Europe's largest producer of organic pest controls. The website is that of the Dutch parent company but there is also an English language version. Information is provided on the company and its distribution, pests and diseases, guidelines, pollination, support and research. Comments are welcomed.
Contact: Julian Ives
Email: info@koppert.co.uk
Address: Homefield Business Park, Homefield Road, Haverhill, Suffolk CB9 8QP
Tel: 01440 704488
Fax: 01440 704487

Lakeland Natural Vegetarian Guest House
www.lakelandnatural.co.uk

Specialities: Guesthouse; holidays; vegetarian
Website: Website provides a house tour, shows the breakfast menu and gives information on the hosts, the locality, activities and booking. The tariff is shown.
Contact: Gerard & Sylvia Daley
Email: relax@lakelandnatural.co.uk
Address: Low Stack, Queen's Road, Kendal, Cumbria LA9 4PH
Tel: 01539 733 011
Fax: 01539 733 011

Lakeland Paints Ltd
www.ecospaints.com

Speciality: Paints
Website: Ecos paints are award winning environmentally friendly, odourless and solvent-free. 84 co-ordinated colours are available. Site gives list of types of paint available and provides an on-line brochure with shade card. There is an email facility for ordering the printed brochure.
Catalogue: Brochure & colour card
Address: Unit 34, Heysham Business Park, Middleton Road, Heysham, Lancashire LA3 3PP
Tel: 01524 858978
Fax: 01524 852371

La Leche League
www.laleche.org.uk

Specialities: Babycare; maternity; breast feeding; organisation
Website: International organisation providing support and information on breast-feeding. Website explains the League and what it does, gives facts and figures and contact information. There is also breastfeeding information, articles and news on events. There is an on-line catalogue of books and leaflets. Membership information is provided.
Contact: Krystina Henderson
Email: lllgb@wsds.co.uk
Address: PO Box 29, West Bridgford, Nottingham, Nottinghamshire NG2 7NP
Tel: 020 7242 1278

Lancrigg Vegetarian Country House Hotel
www.lancrigg.co.uk

Specialities: Hotel; holidays; vegetarian
Website: Hotel in 30 acres of garden overlooking Easedale. Website gives details of the international vegetarian cookery, brief descriptions of the rooms, and the tariff. There is also information on Christmas and the New Year.
Links: Yes
Email: info@lancrigg.co.uk
Address: Easedale, Grasmere, Cumbria LA22 9QN
Tel: 015394 35317
Fax: 015394 35058

Lavera UK
www.lavera.co.uk

Specialities: Bodycare; cosmetics; suncare; vegetarian & vegan; babycare
Website: Extensive range of bodycare and decorative cosmetics. The Neutral range may be used whilst under homoeopathic treatment. On-line there is an animal testing statement and information on suitability for vegetarians and vegans. There is a full product listing with descriptions, ingredients, and advice. Products are available from healthfood and organic shops and also by mail order from the postal address.
Contact: Graeme Hume
Email: trade@lavera.co.uk
Address: PO Box 7466, Castle Douglas, Dumfries & Galloway DG7 2YD Scotland
Tel: 01557 814941
Fax: 01557 814941

Leapingsalmon.com Ltd
www.leapingsalmon.com

Specialities: Ready meals; gourmet home delivery service
Website: Provides all the ingredients in kit form for a gourmet meal for two using only the best and freshest ingredients. Ingredients are sealed in cool bags. Deliveries in London and throughout the UK.
On-line ordering: Yes
Credit cards: Most major cards
Email: james@leapingsalmon.com
Address: Nightingale House, 1 - 7 Fulham High Street, London SW6 3JH
Tel: 0870 701 9100

Leary's Organic Seed Potatoes
www.organicpotatoes.co.uk

Specialities: Seed potatoes; onion sets; garlic sets; wholesaler
Website: Grows and supplies organic seed potatoes. On the website there is a wholesale price list for both the UK and Dutch markets. There is also a wholesale price list for onion and garlic sets.
Contact: Laurence Hasson
Email: lhasson@bigfoot.com
Address: 11 Caledonia Place, Clifton, Bristol BS8 4DJ
Tel: 0117 923 8940
Fax: 0117 973 5158

Lechlade Trout Farm
www.fishlink.com/lechlade

Specialities: Trout; smoked fish
Website: Fresh and smoked Cotswold trout. Website provides a description of the fishery with photographs, details of tickets and price list, competitions and a location map.
Email: tim@timtrout.co.uk
Address: Burford Road, Lechlade, Gloucestershire GL7 3QQ
Tel: 01367 253266
Fax: 01367 252663

The Leela Centre
www.osholeela.co.uk

Specialities: Retreats; workshops
Website: Designed for all forms of personal growth, therapy, healing, meditation, music, dance and theatre, the centre has beautiful group rooms and is set in the countryside. It is run by experienced therapists. Website has contact details and a photo gallery. There is also a calendar and information on training and community experience programmes. Dates of open days which offer a free tour are given.
Links: Yes
Email: info@osholeela.co.uk
Address: Osho Leela,
Thorngrove House,
Common Mead Lane,
Gillingham, Dorset SP8 4RE
Tel: 01747 821 221
Fax: 01747 826 386

Guy Lehmann SFI Ltd
www.guylehmann.com

Speciality: Food industry supplier
Website: Bulk supplies of mustard ingredients, wine, spirits and speciality vinegars. Also supplies organic pumpkin, caraway, linseed, millet and sunflower. Product list on the website lists organic ingredients.
Export: Europe
Contact: Mark Lehmann
Email: sales@guylehmann.com
Address: 23 Glenfield House,
Glenfield Park, Philips Road, Blackburn,
Lancashire BB1 5PF
Tel: 01254 682479
Fax: 01254 682492

Les Jardiniers du Terroir
www.lesjardiniersduterroir.org.uk

Specialities:
Design for organic gardens; box scheme; permaculture
Website: Provides organic and permaculture design and consultancy service. Website was under construction.
Links: Yes
Contact: Richard Smedley
Email:
enquiry@lesjardiniersduterroir.org.uk
Address: 4 Hollins Green, Bradwall,
Middlewich, Cheshire CW10 0LA
Tel: 01270 764960
Fax: 01270 764960

Lessiter's Ltd
www.lessiters.com

Speciality: Chocolate
Website: Makers of fine chocolates since 1911, producing high quality products in the traditional way. On-line catalogue covers the organic collection and other chocolates. There are information pages on organic chocolate. An on-line ordering facility is under development. In the meantime the chocolates are available in some major supermarkets. There is a mailing list for updating information.
Contact: Hans Luder
Email: mail@lessiters.com
Address: 61 London Road,
Woolmer Green, Knebworth,
Hertfordshire SG3 6JE
Tel: 01438 817338
Fax: 01438 817193

Letterbox Library
www.letterboxlibrary.com

Speciality: Children's books
Website: Mail order company specialising in multi-cultural and non-sexist books for children. Books might include children and adults with disabilities, single parent families, children who are adopted or fostered and many more issues and are consequently not easily available on the high street. There is a listing of selected books on the website and a printed catalogue of the full range is available free on request.
Catalogue: Yes
Email: info@letterboxlibrary.com
Address: 71-73 Allen Road,
Stoke Newington, London N16 8RY
Tel: 020 7503 4801
Fax: 020 7503 4800

Little Green Earthlets
www.earthlets.co.uk

Specialities: Babycare; nappies
Website: On-line web shopping for baby-care covering nappies, bedtime, bottom covers, mealtimes, liners, pottytime, playtime, bathtime, health, hygiene, natural clothing, snuggletime and essential extras. There is a page of FAQs largely concerned with the use of washable nappies There is a retail shop in London.
Links: Yes
On-line ordering: Yes
Credit cards: Most major credit cards
Email: sale@earthlets.co.uk
Address: Unit 1 -3, Stream Farm, Chiddingley, Nr Lewes, East Sussex BN8 6HG
Tel: 01825 873301
Fax: 01825 873303

Little Salkeld Watermill
www.organicmill.co.uk

Specialities: Flours; restaurant; bed & breakfast; vegetarian & vegan; cookery courses
Website: Organic flours are milled in traditional fashion using water power in an 18th century watermill. The site gives details of how the mill works, history of the mill, product guide, mail order service and prices, visits to the mill, newsletter, and baking courses. Orders may be placed by phone or email. Baking course details with dates and booking information are also on the site. There are vegetarian and vegan cookery courses.
Contact: Nick Jones
Email: organicflour@aol.com
Address: Little Salkeld, Penrith, Cumbria CA10 1NN
Tel: 01768 881523
Fax: 01768 881047

The Little Herbal Company
www.littleherbal.co.uk

Specialities: African herbal remedies; herbal remedies; skincare
Website: Specialists in traditional African herbal remedies: Themba Herbal skin cream and Simba, the South African potato tuber which is effective as a potential means of immune support. Website gives background and application of the remedies which may be ordered on an email order form.
Credit cards: Visa; Mastercard
Contact: Leslie & Glyn Robinson
Email: enquiries@littleherbal.co.uk
Address: 35 St Georges Road, Scholes, Holmfirth, Huddersfield, West Yorkshire HD7 1UQ
Tel: 01484 685100
Fax: 01484 684954

Living Lightly Limited
www.livinglightly.co.uk

Speciality: Cycle trailers
Website: Website gives information on three models of cycle trailers and luggage carriers with photos and specifications. There is a printable order form with prices.
Catalogue: Yes
Email: info@livinglightly.co.uk
Address: 14 Holly Terrace, York YO10 4DS
Tel: 0800 074 332

Livos Natural Paints
www.livos.com

Speciality: Paints
Website: Manufacturer of natural wood treatments and finishes. Made with pure natural ingredients, based on organic linseed oil. The website is US based. It provides product information and company history but does not give UK prices.
Address: Unit D7, Haws Craft Centre, Jackfield, Ironbridge, Shropshire TF8 7LS
Tel: 01952 883288
Fax: 01952 883200

Lloyd Maunder Ltd
www.meatdirect.co.uk

Specialities: Meat; poultry; ready meals
Website: Nationwide service throughout UK. New internet service which brings top quality fresh meat to the door, including organic beef, lamb and pork, free range poultry and fully prepared gourmet dishes ready for the oven, fridge or freezer. Prices are given on-line. Secure on-line ordering
On-line ordering: Yes
Credit cards: Most major cards
Contact: Richard Maunder
Email: richard@lloydmaunder.co.uk
Address: Willand, Cullompton, Devon EX15 2PJ
Tel: 01884 820534
Fax: 01884 821404

Loch Fyne Oysters Ltd
www.loch-fyne.com

Specialities: Smoked salmon; shellfish
Website: Company undertakes to ensure that the environmental impact of its activities is at least neutral and at best to be positive. It actively works to enhance biodiversity and underpin the economy of the community. Website provides full details on this aspect of the company. It also provides an on-line shop for the product range. There is information on the Loch Fyne Society which is committed to protecting the local ecology.
Links: Yes
On-line ordering: Yes
Credit cards: Visa; Mastercard; Switch; Amex; Diners
Email: sales@loch-fyne.com
Address: Clachan, Caimdow, Argyll PA26 8BL, Scotland
Tel: 01499 600 264
Fax: 01499 600 234

Logona Cosmetics
www.logona.co.uk

Specialities: Bodycare; skincare
Website: Wide range of beauty products which contain organic ingredients wherever possible. The product list on the website covers cosmetics, facial care, hair colours, haircare, bodycare, men's care, hypo-allergenic, oral care and sun care. There is an on-line ordering facility.
On-line ordering: Yes
Credit cards: Most major credit cards
Email: sales@logona.co.uk
Address: Unit 3B, Beck's Green Business Park, Beck's Green Lane, Ilketshall St Andrew, Beccles, Suffolk NR34 8NB
Tel: 01986 781782
Fax: 01449 780297

Lollipop Children's Products
www.teamlollipop.co.uk

Specialities: Baby care; nappies
Website: Babycare products including reusable nappies. Mail order service. On-line ordering was under development at the time of writing but an email order form is provided on the site. Local agents can be contacted throughout much of the UK via the website. Site headings include nappies, bath and bed, wooden toys, out and about, organic for babies, just for mum, environmental matters, and how to order.
Email: liz@teamlollipop.co.uk
Address: Bosigran Farm, Pendeen, Penzance, Cornwall TR20 8YX
Tel: 01736 799512
Fax: 01736 794886

London College of Clinical Hypnosis
www.lcch.co.uk

Specialities: Hypnosis; courses
Website: College was created specifically to provide intending practitioners of hypnosis and hypnotherapy with the skills, disciplines and tuition necessary to practise both soundly and ethically. The website provides information on the prospectus, the courses, regional centre, contacts, free on-line study guide, and details of masterclass topics.
Email: lcch@lcch.co.uk
Address: 15 Connaught Square, London W2 2RL
Tel: 020 7402 9037
Fax: 020 7262 1237

The London College of Massage
www.massagelondon.com

Specialities: Massage; courses; practitioner training
Website: Established to make massage widely available to all and to improve the standard of training throughout the country and position it as a necessary integrated therapy within the world of orthodox medicine. Website gives details of courses and workshops together with a prospectus and also provides information on the clinic and treatments available. A range of quality massage products is listed and may be ordered by phone, fax or post.
Links: Yes
Email: admin@massagelondon.com
Address: 5 Newman Passage, London W1 3PF
Tel: 020 7323 3574
Fax: 020 7367 7125

London College of Shiatsu
www.londoncollegeofshiatsu.com

Specialities: Shiatsu; courses; practitioner training
Website: The college offers various classes and workshops including a three-year practitioner course. It aims to teach and promote meridian and zone therapy as a gentle, yet effective therapeutic style of shiatsu. The website provides details of the courses with term dates, fees, application forms and curriculum. It explains the benefits of shiatsu study and of the diploma. There is also information on the teachers and a prospectus which may be downloaded
Email: info@londoncollegeofshiatsu.com
Address: 25-27 Dalling Road, London W6 0JD
Tel: 020 8741 3323

London College of Traditional Acupuncture & Oriental Medicine
www.lcta.com

Specialities: Acupuncture; oriental medicine; courses; practitioner training
Website: Educates mature students who aspire to the professional practice of oriental medicine in the western world and offers qualifications in holistic healing of the mind, body, emotions and spirit. Website gives background information on acupuncture and Chinese herbal medicine and provides details of professional courses, post-graduate seminars, courses for the interested general public, presentation dates and course dates.
Email: lcta@bogo.co.uk
Address: 1st Floor, HR House, 447 High Road, North Finchley, London N12 0AZ
Tel: 020 8371 0820
Fax: 020 8371 0830

London Farmers' Markets
www.londonfarmersmarkets.com

Speciality: Farmers' markets
Website: Website gives listing of markets in London with locations and travel details. There are also articles on the site, seasonal availability, information on starting a market, recipes and a facility for joining the mailing list.
Links: Yes
Email: info@lfm.demon.co.uk
Address: 6 St Paul's Street, London, N1 7AB
Tel: 020 7704 9659

London Fruit & Herb Company
www.premierbrands.com

Specialities: Fruit teas; herb teas
Website: New innovations in product formulation and recipe development to create a range of fruit and herb teas that taste even better than they smell. All teas are 100% natural with no artificial colours or flavours. Website covers all aspects of tea: history, manufacture, types, ethics, green, organic and fairtrade. There are also comprehensive pages on fruit and herb teas, coffee and malted.
Contact: Gillian Murphy
Email: gillian.murphy@premierbrands.com
Address: Premier Brands (London Fruit & Herb), 9 Kinnaird Park, Newcraighall Road, Edinburgh EH15 3RF, Scotland
Tel: 01506 695340
Fax: 0131 657 4565

Long Distance Walkers Association
www.ldwa.org.uk

Specialities: Activity holidays; organisation; walking
Website: Furthers the interests of those who enjoy long distance walking. Website gives information on the association journal, challenge walks, local groups, long distance paths and outdoor gear bearing the association logo.
Links: Yes
Email: membership@ldwa.org.uk
Address: c/o Bank House, Wrotham, Sevenoaks, Kent TN15 7AE
Tel: 01732 883705

Lotus Emporium
www.lotusemporium.com

Specialities: Incense; aromatherapy products
Website: Incenses are blended from appropriate organic herbs, gums and resins. There is also a wide range of aromatherapy products. All are listed on the web pages and there is a price list. Details of stockists are also provided.
Credit cards: Most major credit cards
Contact: Kate Bartran
Address: Unit 9 Guildhall Market, High Street, Bath BA1 5AQ
Tel: 01225 448011
Fax: 01225 425589

Lunn Links Kitchen Garden
www.kitchen-garden.co.uk

Specialities: Importer; herbs; spices; coffee substitute
Website: Company was formed to act as a link between Third World producers and the buyers in Europe and offer organic products at a fair price to both producers and the buying public. It works closely with an organic farm in Zimbabwe. It markets an organic range of herbs and spices and also sells SoyCaf, a coffee substitute. Website provides company background with pictures and product listing. A list of wholesalers who stock the products is given.
Links: Yes
Contact: Stephen Lunn
Email: llorganic@aol.com
Address: Greenbrier, Victoria Road, Brixham, Devon TQ5 9AR
Tel: 01803 853579
Fax: 01803 883892

Lyme Regis Fine Foods
www.lymeregisfoods.com

Specialities: Snacks; vegetarian & vegan
Website: Healthy snack foods, both organic and non-organic, made from all-natural ingredients, GMO-free, suitable for vegetarians and free from hydrogenated fat. Some special diets are catered for with products suitable for vegan, wheat-free, gluten-free, no-added-sugar and dairy-free diets. All products may be ordered on-line on the website. Organic and other products are listed with illustrations and their ingredients and nutritional information.
Links: Yes
On-line ordering: Yes
Contact: Alaa Zaki
Email: info@lymeregisfoods.com
Address: Unit D, Station Road Industrial Estate, Liphook, Hampshire GU30 7DR
Tel: 01428 722900
Fax: 01428 727222

Lynford Hall Hotel
www.lynfordhallhotel.co.uk

Specialities: Hotel; holidays
Website: Hotel, promoting organic food and wine with conference facilities. Wherever possible fresh seasonal vegetables are used from the hotel's own organic garden and meats come from organic suppliers. There is an extensive range of organic wines, beers and spirits. The website gives illustrated details of the hotel and its restaurant. It also provides information on future developments and directions to find the hotel.
Email: lynfordhallhotel@faxvia.net
Address: Nr Thetford, Norfolk IP25 5HW
Tel: 01842 878351
Fax: 01842 878352

M S B Mastersoil Builders
www.mastersoil.co.uk

Specialities: Fertilisers; gardening; bodycare
Website: Suppliers of organic natural, chemical-free fertilisers, creating gardens which are safer for children, pets and wildlife. Also shampoos and moisturisers. Website provides basic contact information.
Contact: J Warren
Email: contact@mastersoil.co.uk
Address: Unit 3, Binghams Park Farm, Potten End Hill, Waterend, Hemel Hempstead, Hertfordshire HP1 3BN
Tel: 01442 216203
Fax: 01442 268640

Macrobiotic Association of the UK
www.macrobiotics.co.uk

Specialities: Macrobiotics; organisation
Website: There is a history of macrobiotics, definitions and a list of macrobiotic foods. The website includes information on the ten steps to health, yin and yang, cooking tips, library, books, directory and what's new. There is a listing of macrobiotic counsellors.
Links: Yes
Email: info@macrobiotics.co.uk
Address: Vitalise Well-being Services, 30 Forgefield Court, High Street, St Mary Cray, Kent BR5 4AZ
Tel: 01689 896175

Magpie Home Delivery
www.magpiehomedelivery.co.uk

Specialities: Home delivery; recycled household goods; box schemes
Website: Home delivery service in Brighton and Hove area. On-line ordering is available and the site is up-dated regularly, especially the seasonal items.
On-line ordering: Yes
Contact: Daniel
Email: magpiehomedelivery@fastnet.co.uk
Address: Unit 4, Level 3, New England House, New England Street, Brighton, East Sussex BN1 4GH
Tel: 01273 621222
Fax: 01273 626226

The Manchester Cushion Company
www.cushion.org.uk

Specialities: Yoga products; meditation products
Website: Suppliers of cushions and mats for yoga, meditation, massage and shiatsu. Website was under construction at the time of writing.
Catalogue: Brochure available
Email: cushionco@zen.co.uk
Address: 120 Grosvenor Street, Manchester M1 7HL
Tel: 0161 882 0801
Fax: 0161 272 7991

Manor Farm Organic Milk Ltd
www.manor-farm-organic.co.uk

Specialities: Milk; wholesaler
Website: Organic milk from Dorset farms. The website gives contact details and some product information. Milk and cream are available from various retail outlets and from the farm shop.
Contact: Will Best
Email: enquiries@manor-farm-organic.co.uk
Address: Manor Farm, Godmanstone, Dorchester, Dorset DT2 7AH
Tel: 01300 341415
Fax: 01300 341170

Marchents
www.marchents.com

Specialities: Fruit; vegetables; meats; seafood
Website: Offers food from small craft producers with the convenience of a single order point and direct delivery to the customer. Food can be located by category, producer or through a search device and ordered on-line. Catalogues may be ordered on-line.
On-line ordering: Yes
Credit cards: Most major cards
Export: Yes, dependent on locality
Catalogue: Yes
Email: info@marchents.com
Address: Customer Care, 1 Apollo Rise, Southwood, Farnborough, Hampshire GU14 0GT
Tel: 0870 6061624
Fax: 01252 391494

Markus Products Ltd
www.markusproducts.co.uk

Specialities: Butters; cheese;
Website: Manufacturers of garlic and savoury butters and spreads supplied chilled or frozen in the UK to caterers, wholesalers, restaurants and similar outlets. Website provides listing and details of products and services available. Products are tailored to customer specifications.
Contact: Simon Clarke
Email: simon@markusproducts.co.uk
Address: 19 Old Market Centre, Station Road, Gillingham, Dorset SP8 4QQ
Tel: 01747 823716
Fax: 01747 825692

Marlborough House
www.marlborough-house.net

Specialities: Guesthouse; vegetarian; holidays
Website: A small hotel in the heart of Georgian Bath especially noted for its eclectic cuisine using fine fresh ingredients, organic and GM-free. Website provides description of the hotel, its cuisine and its rooms as well as information on Bath. Room prices are provided
Credit cards: Visa; Mastercard; Amex; Switch; Diners
Contact: Laura & Charles Dunlap
Email: mars@manque.dircon.co.uk
Address: 1 Marlborough Lane, Bath BA1 2NQ
Tel: 01225 318 175
Fax: 01225 466 127

Martin Pitt Freedom Eggs
www.freedomeggs.co.uk

Speciality: Eggs
Website: Genuine free range and organic eggs from happy hens on a real farm. Website provides background information on egg production. Eggs are free from colourants, chemicals and hormones. Details of stockists may be obtained by phoning. There is no mail order but a location map is provided on the website.
Contact: Martin Pitt
Email: gerry.tuffs@virgin.net
Address: Great House Farm, Gwehelog, Nr Usk, Monmouthshire NP15 1RJ, Wales
Tel: 01291 673129
Fax: 01291 672 766

Mastic Gum Europe Ltd
www.mastika.com

Specialities: Mastic gum; herbal remedies
Website: The exclusive European based supplier of mastic gum based health care products. Mastic gum has traditionally been used for over 3,000 years as a natural digestive aid and may help to maintain a healthy digestion. The website has both a consumer and trade section. The consumer site gives information on mastic gum in general as well as on specific products.
Contact: Farhad Alaaldin
Email: enquiries@mastika.com
Address: 344 Kilburn High Road, London NW6 2QJ
Tel: 0870 740 3850
Fax: 0870 740 3860

F W P Matthews
www.fwpmatthews.co.uk

Speciality: Flours
Website: Wholemeal flours, whole grain and wheatfeed. Website has a page giving details of organic flours. A mail order catalogue is available and an order form can be downloaded from the site. Prices include carriage.
Links: Yes
Credit cards: Most major cards
Catalogue: Yes
Contact: Paul Matthews
Email: sales@fwpmatthews.co.uk
Address: Station Road, Shipton under Wychwood, Chipping Norton, Oxfordshire OX7 6BH
Tel: 01993 830342
Fax: 01993 831615

Maui Noni
www.mauinonijuice.com

Specialities: Noni juice; food supplement
Website: Maui Noni is the only pharmaceutically patented Noni product in the world. Noni is a tropical fruit from the islands of the Pacific Ocean. It is reputed to have therapeutic qualities and information on clinical studies is provided. A full range of Noni products is made available in the UK: juice, capsules, health bars and bodycare products.
Contact: Trevor Partridge
Email: enquiries@mauinonijuice.com
Address: Anncome Ltd, Laurel Grange, Whitmore Vale, Grayshott, Hindhead, Surrey GU26 6LE
Tel: 01428 606625
Fax: 01428 608197

www.mavcohealth.com

Specialities: Herbal remedies; sports nutrition
Website: Percutane Natural Cream is a New Zealand, clinically developed, topical and dual-purpose remedy containing sports and health science with a unique formulation of natural plants and marine extracts for the relief of arthritic discomfort, aching muscles and joints and as a pre-activity warm up. Website gives product information.
Links: Yes
Contact: Martin Burns
Email: martinnb@mavcohealth.com
Address: PO Box 25057, London SW4 0WY
Tel: 0788 770 1824
Fax: 0207 622 4776

N & J Mawson
www.sarsaparilla.co.uk

Specialities: Juices; cordials
Website: Traditional juices and cordials, produced to the original recipe. Sarsaparilla cordial is made from natural ingredients including ginger, liquorice and sarsaparilla. There is an on-line order form as well as a list of stockists indexed by postcode,
On-line ordering: Yes
Credit cards: Most major credit cards
Contact: Nigel Mawson
Email: info@sarsaparilla.co.uk
Address: 57 George Street, Oldham, Lancashire OL1 1LT
Tel: 0161 626 0341
Fax: 0161 642 8182

Maximuscle
www.maximuscle.com

Specialities: Sports nutrition; food supplements; magazine
Website: Europe's largest sports nutrition company. An extensive site filled with information. Supplements are listed on the web pages and can be ordered publisher of the magazine *Muscular Performance.*
On-line ordering: Yes
Credit cards: Most major credit cards
Contact: Jeannette Clarke
Email: info@maximuscle.com
Address: 40 Caxton Way, Watford Business Park, Watford, Hertfordshire WD18 8JZ
Tel: 01923 650600

Maya Books
www.mayabooks.ndirect.co.uk

Speciality: Books
Website: On-line catalogue of books covering agriculture, building, directories, gardening, politics, renewable energy, sustainable architecture, waste management, and water and sewage solutions. On-line secure ordering.
Links: Yes
On-line ordering: Yes
Credit cards: Visa; Mastercard; Switch; Delta; Solo
Export: Yes
Email: sales@mayabooks.ndirect.co.uk
Address: PO Box 379, Twickenham, Middlesex TW1 2SU
Tel: 0208 287 9068

Mayfield Services
www.mayfield-excalibur.co.uk

Speciality: Food dehydrators
Website: Food dehydrators imported from the USA which can be used for almost any fruit or vegetables to produce dried foods which are 100% free of chemicals, preservatives or added sugar. Website provides information on the different products available with prices, illustrations and specifications. There is also a page of ideas and suggestions for their use.
Contact: Richard Hobbs
Email: mail@mayfieldservices.fsnet.co.uk
Address: PO Box 20124, Kenilworth, Warwickshire CV8 2WP
Tel: 01926 854443
Fax: 02476 688603

www.meatmatters.uk.com

Specialities: Meat; fish; dairy; poultry; fruit; vegetables
Website: Suppliers of fresh natural produce: organic meat, poultry, eggs, fish, fruit, vegetables and groceries. Sampler pack available. Free delivery in M25 area, Hertfordshire, South Buckinghamshire, Berkshire and Oxfordshire. Small charge elsewhere. Secure on-line ordering. There is an information page on organic meat.
On-line ordering: Yes
Credit cards: Most major credit cards
Catalogue: Price list
Email: enquiries@meatmatters.uk.com
Address: 2 Blandys Farm Cottages, Letcombe Regis, Wantage, Oxfordshire OX12 9LJ
Tel: 08080 067426
Fax: 01235 772526

Medihoney
www.medihoney.com

Speciality: Honey
Website: Website gives details of active therapeutic honey for dietary use and oral health. Also sterilised high potency antibacterial honey for wound care. There are information pages with academic references on the usage of honey in a medical context. There are sections on facts, research and reference. Medihoney is available through pharmacies but the site gives a link for international on-line purchasing.
Links: Yes
Email: info@medihoney.com

Meridian Foods
www.meridianfoods.co.uk

Speciality: Sauces
Website: Organic, additive-free products. The mission is to produce healthy foods which are convenient and value for money. On the website there is an extensive organic information section. There are also product listings and a worldwide.
Links: Yes
Email: info@meridianfoods.co.uk
Address: Unit 13, WDA Advance Factories, Corwen, Denbighshire LL21 9RJ, Wales
Tel: 01490 413151
Fax: 01490 412032

The Mexican Hammock Co
www.hammocks.co.uk

Specialities: Hammocks; Mexican goods
Website: Established to provide an outlet in Europe for co-operative based groups in rural Mexico producing hand-woven Mayan hammocks. Company is a joint venture with a small group of villages with the aim of developing reliable income in areas of high unemployment and relative poverty. Website provides general information on the care and use of hammocks as well as an on-line catalogue with ordering facility. It also sells garden furniture, mirrors, rugs, pottery and blankets.
On-line ordering: Yes
Credit cards: Visa; Mastercard; Switch; Amex
Email: hammocks@clara.co.uk
Address: 42 Hill Avenue, Victoria Park, Bristol BS3 4SR
Tel: 0117 972 4234
Fax: 0117 972 4234

Miniscoff
www.miniscoff.co.uk

Specialities: Children's food; ready meals for children
Website: Organic food for children. Fresh, organic and complete meals for toddlers and children prepared with the best ingredients. Website provides a listing of the meals available and details of stockists. There is a feedback email form with a questionnaire.
Email: goodfood@miniscoff.co.uk
Address: Ham House, Ham Green, Holt, Wiltshire BA14 6PY
Tel: 020 7731 1605

MINTON HOUSE
www.mintonhouse.co.uk

Specialities: Retreats; conference facilities
Website: Overlooking Findhorn Bay, Minton House is a retreat and education centre and a place for respite care. It may be rented by groups for small conferences and workshops. The website provides details and a programme of retreats and workshops, summer music courses and bed and breakfast facilities.
Links: Yes
Credit cards: Visa; MasterCard
Contact: Jules
Email: minton@findhorn.org
Address: Findhorn Bay, Findhorn, Forres, Moray IV36 3YY, Scotland
Tel: 01309 690 819
Fax: 01309 691 583

MONTEZUMA'S CHOCOLATES
www.montezumas.co.uk

Specialities: Chocolate; vegan
Website: Montezuma's sell organic chocolate that is suitable for vegans and is free of genetically modified ingredients. Colourful website provides background information on chocolate production including details for vegans. There is an on-line chocolate shop with illustrated and descriptive catalogue. There is also information on stockists.
Links: Yes
On-line ordering: Yes
Credit cards: Most major cards
Email: simon.pattinson@montezumas.co.uk
Address: 15 Duke Street, Brighton, East Sussex BN1 1AH
Tel: 01273 324 979
Fax: 01273 711 710

MOTHER HEMP
www.motherhemp.com

Specialities: Fabrics; clothes; hemp goods; ice cream
Website: Mail order organic fabrics and clothes. The company was set up for the advancement of hemp farming, processing and distribution with environmental issues central to its agenda. The site has pages on ecology, history, cultivation, seed, oil, and fabrics. There is an on-line catalogue for product ordering.
Links: Yes
On-line ordering: Yes
Credit cards: All major cards
Catalogue: Free catalogue
Email: info@motherhemp.com
Address: Tilton Barns, Tilton Lane, Nr Firle, Lewes, East Sussex BN8 6LL
Tel: 07041 31 32 330
Fax: 020 7691 7475

MOTHERNATURE
www.mothernaturebras.co.uk

Specialities: Maternity; aromatherapy products
Website: A small specialist company selling nursing bras and maternity support garments. There is a page giving detailed advice on sizing and measuring. Full details and illustrations are given for the products.
On-line ordering: Yes
Credit cards: Most major credit cards
Catalogue: Brochure available
Contact: Felicity Tucker
Email: felicity@mothernaturebras.co.uk
Address: Acorn House, Brixham Avenue, Cheadle Hulme, Cheshire SK8 6JG
Tel: 0161 485 7359
Fax: 0161 485 7559

Mountain Buggy
www.mountainbuggy.com

Speciality: Baby buggies
Website: Baby buggies designed for rugged quality, safety and innovative designs. The design for the buggies originated in New Zealand. This is an international website with a UK section. All the buggy models are illustrated with details of specifications and accessories available. There is information on safety and awards and there is a photo gallery. A regional listing of UK retail outlets with contact numbers is provided.
Email: mountainbuggyuk@cs.com
Address: 9 Cromwell Road, Camberley, Surrey GU15 4HY
Tel: 01276 502 587
Fax: 01276 502 587

Mr Fothergill's Seeds
www.mr-fothergills.co.uk

Specialities: Seeds; vegetable seeds; gardening
Website: The range includes organic, wildflower and heritage seeds. There is an email request for ordering a printed catalogue. Website was under construction at the time of writing.
On-line ordering: Yes
Export: Yes
Catalogue: Yes
Email: info@mr-fothergills.co.uk
Address: Gazely Road, Kentford, Newmarket, Suffolk CB8 7QB
Tel: 01638 751161
Fax: 01638 751624

Mrs Moons
www.mrsmoons.com

Specialities: Baking mixes; cakes; vegetarian
Website: Producers of organic baking mixes. Also specialist organic bakers of cakes, muffins and cookies. Website gives background history of the company and illustrated details of the mixes. Mail order is available for direct customers and a list of wholesale stockists is given for retailers. Recipes and baking tips are given.
Contact: Tony Davies
Email: mrsm@mrsmoons.com
Address: Walton Farm, Kilmersdon, Somerset BA3 5SX
Tel: 01761 432382
Fax: 01761 439645

Mrs Tee's Wild Mushrooms
www.wildmushrooms.co.uk

Specialities: Mushrooms; cookery courses
Website: Wild mushrooms supplied both to restaurants and to individual customers. Local groups can buy mushrooms at wholesale prices. There are also residential cookery courses. The website provides information on retail prices, accommodation in the guest house with illustrations, and details of regular seminars.
Email: mrs.tees@wildmushrooms.co.uk
Address: Gorse Meadow, Sway Road, Lymington, Hampshire SO41 8LR
Tel: 01590 673594
Fax: 01590 673376

Mufti
www.mufti.co.uk

Speciality: Furniture
Website: Luxury hand made furniture using natural raw materials, reclaimed wood and recycled cotton. Some items may be ordered on-line.
On-line ordering: Yes
Contact: Michael D'Souza
Email: info@mufti.co.uk
Address: 789 Fulham Road, London SW6 5HA
Tel: 020 7610 9123
Fax: 020 7384 2050

Munson's Poultry
www.munsonspoultry.demon.co.uk

Specialities: Poultry; turkeys; meat
Website: Free roaming birds whose diet is supplemented by corn. Supplies through a chain of selected prestige retail outlets throughout the UK. A list is given on the website. Also mail order form can be printed from the site and prices are given. There is a page of photos of the poultry, ducks, turkeys and sheep.
Credit cards: Most major cards
Address: Emdon, Straight Road, Boxted, Colchester, Essex CO4 5QX
Tel: 01206 272637
Fax: 01206 272962

Peter & Therese Muskus
www.bigfoot.co./muskus

Specialities: Chalet; caravan; tipi; holidays
Website: Chalet, caravan and tipi with beautiful outlook over loch and woods. Roe deer and osprey watching. Website gives full details of the different types of accommodation with accompanying layouts and photographs. Prices and conditions are listed. WWOOFers are welcomed.
Links: Yes
Contact: Peter & Therese Muskus
Email: muskus@bigfoot.com
Address: Laikenbuie, Nairn, Highland IV12 5QN, Scotland
Tel: 01667 454630

Mycology Research Laboratories
www.mycologyresearch.com

Specialities:
Mushroom nutrition products; herbal remedies
Website: MRL is a global leader in the marketing of mushroom nutrition products. Its proprietary Japanese cultivation technology consistently produces uniform, contaminate-free mushroom powder which is then manufactured into tablets. Website gives details of products with fact sheets, cultivation and manufacturing processes as well as packaging and prices. There are also pages on animal health and research and development and also *Mycology News*.
Contact: William Ahern
Email: info@aneid.pt
Address: Bank House, Saltground Road, Brough, East Yorkshire HU15 1EF
Tel: 01482 667634
Fax: 01482 667859

Myriad Organics
www.myriadorganics.co.uk

Specialities: Meat; vegetables; fruit; bread; household products
Website: One-stop organic shop stocking complete range of organic foods and drinks. Much is locally produced including vegetables from their own market garden. There is a general product listing available on-line.
On-line ordering: No
Contact: Jane Straker
Address: 22 Corve Street, Ludlow, Shropshire SY8 1DA
Tel: 01584 872665
Fax: 01584 879356

Nairn's Oatcakes
www.simmers-nairns.co.uk

Speciality: Oatcakes
Website: The range includes products that are organic, wheat-free, sugar-free and high in fibre. Website gives details of biscuits and oatcakes and their history. There is a contact page for trade enquiries.
Contact: Sharon Robertson
Email: info@simmersofedinburgh.co.uk
Address: 90 Peffermill Road, Edinburgh, Lothian EH16 5UU, Scotland
Tel: 0131 620 7000
Fax: 0131 620 7750

Napier Brown & Co Ltd
www.napierbrown.co.uk

Specialities: Sugars; syrups
Website: The largest independent sugar distributor in the UK. Supplies organic sugars and syrups in retail and industrial packs. There is an on-line illustrated brochure. There is also an industrial ingredient price list and a retail and catering price list on the site.
Contact: C Seargant
Email: sales@napierbrown.co.uk
Address: 1 St Katharine's Way, London E1W 1XB
Tel: 020 7335 2500
Fax: 020 7335 2502

Napiers Herbs
www.napiersherbs.co.uk

Specialities: Herbs; farm shop; courses; alternative medicine; vegetarian & vegan
Website: Provides an holistic base that incorporates herb plants with natural healing. Treatments include reiki, seichim, flower essence remedies, numerology and naturopathy. The farm shops sells organically grown fruit and vegetables suitable for vegans and vegetarians. Website gives details and prices of various treatments. There is information on courses.
Contact: Liz Adams
Email: lizadams@napiersherbs.co.uk
Address: Colchester Road, Tiptree, Essex CO5 0EX
Tel: 01621 815238

The Nappy Lady
www.thenappylady.co.uk

Speciality: Nappies
Website: This very comprehensive website explains the benefits of real nappies and discusses the choices available. It provides tips, birth stories and comments. The on-line store provides a full price list, product information, favourite things, spreading the cost and ordering information.
On-line ordering: Yes
Credit cards: All major credit cards
Contact: Morag Gaherty
Email: gaherty@bigfoot.com
Address: Arcady, 16 Hill Brow, Bearstead, Maidstone, Kent ME14 4AW
Tel: 01622 739 034

The Nappy Shop
www.thenappyshop.co.uk

Specialities: Nappies; babycare
Website: Informative website which provides cost comparisons between the use of different sorts of nappies. Also gives links to research on health. nappies and full explanation of what is required in using them. On-line shopping facility.
Links: Yes
On-line ordering: Yes
Address: Hippocampus, Glengar, Frome Park Road, Stroud, Gloucestershire GL5 3LF
Tel: 01453 756219

National Association of Environmental Education (UK)
www.naee.co.uk

Specialities: Education; environment; organisation
Website: Association of head teachers, teachers, lecturers, advisers, inspectors and all others concerned with the environment. It aims to promote environmental education in all phases of formal education and in other sections of society. The website defines the objectives, summarises the curriculum strategy and provides a bibliography and publications list, details of other organisations.
Links: Yes
Email: mail@naee.co.uk
Address: University of Wolverhampton, Walsall Campus, Gorway Road, Walsall, West Midlands WS1 3BD
Tel: 01922 631 200
Fax: 01922 631 200

National Association of Nappy Services
www.changeanappy.co.uk

Specialities: Organisation; nappies; babycare
Website: A non profit making organisation which sets out the industry's operating standards and guidelines. The website provides a regional facility for finding a local nappy service. There is also an FAQ section on the use of nappy services and washable nappies.
Links: Yes
Address: Unit 1, Hall Farm, South Moreton, Oxfordshire OX11 9AH
Tel: 0121 693 4949

National Energy Foundation
www.greenenergy.org.uk

Specialities: Organisation; energy; alternative energy
Website: Provides information on energy efficiency and renewables. It also undertakes research and training activities and has an educational programme with information and activities for schools. Website headings include projects, renewable energy, green electricity, suppliers, training and news.
Links: Yes
Email: renewables@greenenergy.org.uk
Address: The Energy Centre, Davy Avenue, Knowlhill Road, Milton Keynes, Buckinghamshire MK5 8NG
Tel: 01908 665555
Fax: 01908 665577

National Farmers' Market Association
www.farmersmarkets.net

Specialities: Organisation; farmers' markets
Website: Association to promote existing markets and assist in the formation of new ones. Full listing of farmers' markets in the UK and information on setting up a farmers market is available (send large SAE). There is a similar listing on the website which may be searched by county. There are pages of news and views and getting started information.
Links: Yes
Email: nafm@farmersmarkets.net
Address: South Vaults, Green Park Station, Green Park Road, Bath BA1 1JB
Tel: 01225 787914
Fax: 01225 460840

National Federation of Spiritual Healers (NFSH)
www.nfsh.org.uk

Specialities: Spiritual healing; organisation
Website: Website explains the nature of spiritual healing through the channeling of energies by the healer to re-energise the patient to deal with illness or injury. It provides information on the NFSH and what it does and answers questions. A database is maintained by the federation for healers who are willing to make and receive visits and distance healing is also available.
Email: office@nfsh.org.uk
Address: Old Manor Farm Studio, Church Street, Sunbury-on-Thames TW16 6RG
Tel: 01932 783 164
Fax: 01932 779 648

National Institute of Medical Herbalists
www.btinternet.com/~nimh

Specialities: Organisation; herbal remedies
Website: Professional body of practising herbalists. Provides information on all aspects of western herbal medicine and advises on how to locate a qualified practitioner. Website includes sections on education and training, politics, FAQs, herbal teas and herbs in general.
Links: Yes
Email: nimh@ukexeter.freeserve.co.uk
Address: 56 Longbrook Street, Exeter, Devon EX1 6AH
Tel: 01392 426022
Fax: 01392 498963

National Recycling Forum
www.nrf.org.uk

Specialities: Organisation; recycling
Website: Group of organisations which aims to co-ordinate the development of recycling and waste minimisation in the UK. It acts as a catalyst for new ideas. A Recycled Products Guide is provided in full on the website. The Buy Recycled website has a guide to recycled products, companies and brand names with a searchable database. This is also available as a hardback directory, the *UK Recycled Products Guide.*
Email: nrf@wastewatch.org.uk
Address: NRF Administration, Europa House, 13 - 17 Ironmonger Row, London EC1V 3QG
Tel: 020 7253 6266
Fax: 020 7253 5962

Natrahealth
www.natrahealth.com

Speciality: Herbal remedies
Website: An emerging force in the nutraceutical/botanical field, this British company is committed to developing innovative products from natural sources. Recently launched is a unique herbal balm for joint conditions. Web pages give information on innovative, natural healthcare alternatives as well as contact details and where to purchase the products.
Contact: Barry Smith
Email: karen@natralabs.com
Address: Nutralife(UK) Ltd, Omicron House, Fircroft Way, Kent TN8 6EL
Tel: 01732 866686
Fax: 01732 866718

The Natural Baby Company
www.naturalbabycompany.com

Speciality: Nappies
Website: Company sell two kinds of eco-friendly disposable nappies by mail order. Orders can be placed by post, phone or email.
Links: Yes
Credit cards: Most major credit cards
Email: info@naturalbabycompany.com
Address: 63 Queenstown Road, London SW8 3RG
Tel: 020 7498 9490
Fax: 0870 054 3757

Natural Building Technologies Ltd
www.natural-building.co.uk

Specialities: Paints; reed and clay boards; insulation; building materials
Website: All products are made from eco-friendly and sustainable materials: natural paints, pigments and colour concentrates; earth plasters, daub and lime putty, reed and clay boards, reed rolls, wool and cellulose insulation. Also paint and varnish, graffiti and artex removers. Website gives information on the company and its products, on ecology in general and how to buy. There are also technical pages.
Email: info@natural-building.co.uk
Address: Cholsey Grange, Ibstone, High Wycombe, Buckinghamshire HP14 3XT
Tel: 01491 638911
Fax: 01491 638630

Natural By Nature Oils Ltd
www.naturalbynature.co.uk

Speciality: Aromatherapy products
Website: 27 years hand-on experience in the forefront of essential oils. The website was under construction at the time of writing.
Contact: Franzesca Watson
Email: natbynat@btinternet.com
Address: The Aromatherapy Centre, 9 Vivian Avenue, Hendon Central, London NW4 3UT
Tel: 01582 840848
Fax: 01582 840987

Natural Collection
www.naturalcollection.com

Specialities: Household; babywear; gardening; clothing; vegetarian and vegan; alternative energy
Website: Extensive mail order catalogue with a wide environmental product range covering the organic garden, the natural home, smart energy, and organic clothing and babywear. Natural Collection acts as a trading partner of Friends of the Earth UK.
On-line ordering: Yes
Credit cards: Most major cards
Export: Yes
Catalogue: Yes
Email: sales@naturalcollection.com
Address: Eco House, Monmouth Place, Bath, BA1 2DQ
Tel: 0870 331 333
Fax: 01225 469673

The Natural Cookery School
www.montsebradford.com

Specialities: Cookery courses; vegetarian
Website: Cookery course covers healthy recipes and develops into a comprehensive study of the energetics of foods, the healing power of cooking and the transformative qualities of a natural lifestyle. Website provides course information with full programme details, venues and costs. There is also a monthly recipe page with menus and recipes. Courses also take place in Spain.
Links: Yes
Contact: Montse Bradford
Address: Greenfields, Lovington, Castle Cary, Somerset BA7 7PX
Tel: 01962 240 641
Fax: 01962 240 229

The Natural Death Centre
www.naturaldeath.org.uk

Specialities: Natural death; organisation
Website: The website gives details of the organisation, its activities and its publications including the new *Natural Death Handbook* which details woodland burial grounds in the UK. For a donation the organisation will email information on how to organise a woodland or inexpensive funeral.
Email: rhino@dial.pipex.com
Address: The Natural Death Centre, 20 Heber Road, London NW2 6AA
Tel: 020 8208 2853

Natural Eco Trading Ltd
www.greenbrands.co.uk

Specialities: Cleaning products; household products
Website: The philosophy is simple - use of natural, replenishable ingredients which provide effective cleaning power and good value. Ingredients are renewable, plant- based and natural with no phosphates, chlorine, ammonia or other harsh chemicals and have not been tested on animals. On-line shop covers the kitchen, bathroom, general cleaning, aromatherapy air freshener and paper products. There is a section of testimonials.
On-line ordering: Yes
Credit cards: Most major credit cards
Email: earthfriendly@netrading.freeserve.co.uk
Address: Park Lodge, Court Road, Tunbridge Wells, Kent TN4 8EB
Tel: 01892 616 871
Fax: 01892 616 238

Natural Flooring Direct
www.naturalflooringdirect.com

Speciality: Flooring
Website: Importers and installers of hard wood and natural flooring including seagrass, wool, jute and sisal. Website was under construction at the time of writing.
Email: nfd@eidosnet.co.uk
Address: 46 Webbs Road, Battersea, London SW11 6SF
Tel: 0800 454 721

Natural Friends
www.natural-friends.com

Speciality: Introduction agency
Website: Unique introduction agency for all intelligent, non-smoking, environmentally-concerned singles with a social conscience. Website provides reasons to join, the type of person who joins, how to join and the costs. There is a form for joining on-line.
Links: Yes
On-line ordering: Yes
Credit cards: Visa; Mastercard; Delta; Solo; Switch
Email: info@natural-friends.co.uk
Address: 15 Benyon Gardens, Culford, Bury St Edmunds, Suffolk IP28 6EA
Tel: 01284 728 315
Fax: 01284 728 315

Natural Instincts
www.natural-instincts.com

Specialities: Clothing; fabrics
Website: Mail order available for clothing and household textiles made from organic cotton, wool and linen and also organic knitwear. On-line shop.
On-line ordering: Yes
Credit cards: Visa; Mastercard
Email: doogan@iol.ie
Address: Kilcar, Co Donegal, Ireland
Tel: 00 353 733 8256
Fax: 00 353 733 8258

Natural Science.com Ltd
www.lice.co.uk

Specialities: Headlice; haircare
Website: Head lice treatment which contains no toxic ingredients and is based on neem oil from India. Full instructions and details are provided. Company also sells head lice combs and echinacea drops. There is a page of customer testimonials.
On-line ordering: Yes
Credit cards: Most major cards
Contact: Shaun Hopkins
Email: shaun.hopkins@btconnect.com
Address: 2 Lindsdale House, Middle Street, Llandrindod Wells, Powys LD1 5ET, Wales
Tel: 01597 823 964
Fax: 01597 825 215

Natural Surroundings
www.hartlana.co.uk /natural/wfs.htm

Specialities: Seeds; wildflowers; gardening
Website: Selected on-line catalogue of plants, bulbs, seeds, mixtures and related products. There is a catalogue request form. Services include new habitat creation: ponds, wildflower meadows, hedgerows, woods, copses and butterfly habitats.
Catalogue: Yes
Email: loosley@farmersweekly.net
Address: Centre for Wildlife Gardening and Conservation, Bayfield, Norfolk NR 25 7JN
Tel: 01263 711091

Natural Wheat Bag Company
www.thenaturalwheatbag.co.uk

Specialities: Wheat bags; heat therapy
Website: Suppliers of grain filled products. Wheat bags can be heated and then placed on the distressed area of the body for instant soothing, warmth and relaxation. They can also be frozen and used as a cold compress. Wholesale prices on request. Website gives general information only.
Email: sales@thenaturalwheatbag.co.uk
Address: Unit 10 - 14, Emley Moor Bus Park, Leys Lane, Emley, Huddersfield, Yorkshire HD8 9QY
Tel: 01924 849651
Fax: 01924 840648

Natural Woman
www.natural-woman.co.uk

Specialities: Sanitary products; tampons; babycare
Website: Advice and natural and alternative products for people who care about their health, their families and the environment. Organic tampons and menstrual pads, babycare, vitamin and mineral supplements. There are pages of health information on the site with a key word search facility. Orders can be phoned or faxed on a printable order form.
Links: Yes
Credit cards: Most major credit cards
Export: Yes
Email: sales@natural-woman.com
Address: Freepost (SWB 229), Clifton, Bristol BS8 2ZZ
Tel: 0117 946 6649
Fax: 0117 970 6988

Naturemade
www.naturemade.co.uk

Specialities: Goat dairy products; sheep dairy products; vegetarian
Website: Goat's and sheep's milk and allied products and organic vegetarian foods. Website gives details of the mail order service and an on-line ordering facility is under development.
On-line ordering: Yes
Contact: Esme Brown
Email: sales@naturemade.co.uk
Address: East Johnstone, Bish Mill, South Molton, Devon EX36 3QE
Tel: 01769 573571
Fax: 01769 573571

Natures Cocoons
www.naturescocoons.co.uk

Specialities: Hampers for children; homoeopathy; aromatherapy products; flower remedies
Website: Helping the parent to help the child to help the earth. The company offers hampers that bring together homoeopathy, flower remedies, natural healthcare products and organic food. Hampers are available for new-born babies, weaning babies, crawlers, toddlers, pre-school and older children or they can be made to order. There is also an alternative first aid kit.
Email: info@naturescocoons.co.uk
Address: Galamar, 17 Rodborough Avenue, Stroud, Gloucestershire GL5 3RR
Tel: 01453 767 973

Nature's Dream
www.naturesdream.co.uk

Specialities: Bodycare; haircare; food supplements; insect repellent
Website: Importers of Naturtint, a hypoallergenic, permanent home colorant free from harsh ingredients and enriched with plant extracts. Web pages include customer information centre with suggested precautions and advice before use. There is also a product listing.
Contact: Janine Carroll
Email: sales@naturesdream.co.uk
Address: 41 Regent Street, Rugby, Warwickshire CV21 2PE
Tel: 01788 579957
Fax: 01788 579953

Nature's Plus
www.naturesplus.com

Specialities: Food supplements; vitamins and minerals; bodycare
Website: Manufacturer of over 900 products including vitamins, minerals, standardised herbs, protein and energy shakes and bodycare products. On the website there is a find facility by structure and function category and also by keyword and nutrition category. There is a new product page. The on-line health library contains useful articles which can also be searched by keyword. Local stockists can be located by a postal code search facility.
Contact: Steve Inglin
Email: uksales@naturesplus.com
Address: 12 Harley Street, London W16 9PG
Tel: 0800 917 9030
Fax: 0800 096 4831

Natures Store Ltd
www.naturesstore.co.uk

Specialities: Wholesaler; wholefoods
Website: Over 7,500 wholefood and natural products, many organic, wholesaled and distributed to the independent healthfood trade. Extensive information about the company and its organisation is given on the website. Product details are provided.
Contact: Richard Starkey
Email: health@naturesstore.co.uk
Address: Unit 2, Jamage Industrial Estate, Talke, Stoke-on-Trent, Staffordshire ST7 1XN
Tel: 01782 794300
Fax: 01782 774698

Nature's Treasures Ltd
www.aromatherapy.ndirect.co.uk

Speciality: Aromatherapy products
Website: Oils are fresh, of high quality and sourced from all over the world. Oils can be purchased singly or in blends, and in pure form or in base carriers. Website provides a products and price listing. There is a printable order form. Information is given for therapists and students.
Credit cards: Visa; Mastercard; Access; Amex
Catalogue: Catalogue
Email: naturestreasures@ndirect.co.uk
Address: Bridge Industrial Estate, New Portreath Road, Bridge, Cornwall TR16 4QL
Tel: 01209 843 881
Fax: 01209 843 882

Naturesave Policies
www.naturesave.co.uk

Specialities: Insurance; financial advice
Website: Ethical insurance to protect the customer and the environment. 10% of premiums on household and travel policies go to benefit environmental and conservationist organisations. Website gives mission statement and explanation of the ethos of the company. There is a downloadable series of forms for quotations and claims. There is also information on commercial insurance.
Links: Yes
Email: mail@naturesave.co.uk
Address: Freepost SWB30837, Totnes, Devon TQ9 5ZZ
Tel: 01803 864 390
Fax: 01803 864 441

Natureworks
www.natureworks.co.uk

Speciality: Complementary medicine
Website: One of the first dedicated holistic therapy centres in London. The aim of the clinic is to promote the overall well-being of the individual through natural therapies and a comprehensive range of disciplines offered. The staff can offer guidance and advice in choosing the right therapy. Website gives information on the facilities, the therapies and the therapists.
Email: info@natureworks.co.uk
Address: 16 Balderton Street, London W1Y 1TF
Tel: 020 7355 4036
Fax: 020 7629 2809

Naturex
www.naturex.co.uk

Specialities: Herbal teas; herbal remedies; homoeopathy; aromatherapy products; bodycare
Website: Therapeutic herbal tea formulas based on traditional Mediterranean recipes developed over hundreds of years. A wide range of other product categories is also featured on the extensive website. There is an on-line ordering facility for the product ranges. The website was largely under construction at time of writing.
On-line ordering: Yes
Credit cards: Most major credit cards
Contact: Goran Sumkoski
Email: sales@naturex.co.uk
Address: 62 The Broadway, Tolworth, Surrey KT6 7HR
Tel: 020 8399 3932
Fax: 020 8399 3932

Neal's Yard Agency for Personal Development
www.nealsyardagency.com

Speciality: Holidays
Website: Information and advice on a whole range of holistic and healthy holidays in the UK and overseas. The website was under construction at the time of writing.
Catalogue: Free events guide
Email: info@nealsyardagency.com
Address: BCM Neal's Yard, London WC1N 3XX
Tel: 0870 444 2702
Fax: 0870 444 2702

Neal's Remedies
www.nealsyardremedies.com

Specialities: Herbal remedies; aromatherapy products; homoeopathy; toiletries; courses; books
Website: A wide range of organic products. Neal's Yard is a manufacturer, wholesaler and retailer of natural toiletries, essential oils, homoeopathic remedies and herbal products. It has 20 retail outlets internationally and operates a UK based mail order service and website; it also sells its toiletries to selected trade accounts. The product pages on the web site are conveniently indexed by usage and there is also a product search facility. There is a health information section. There is a find facility for stockists of the products in the UK. Courses on aromatherapy and natural medicine are listed.
On-line ordering: Yes
Credit cards: Most major credit cards
Contact: Sharon Lock
Email: mail@nealsyardremedies.com
Address: 26 - 31 Ingate Place, Battersea, London SW8 3NS
Tel: 020 7498 1686
Fax: 020 7498 2505

Network of European World Shops
www.worldshops.org

Specialities: Organisation; fairtrade
Website: NEWS exists to initiate, direct and promote joint campaigns and co-ordinate Europe-wide activities to promote fairtrade in general and the World Shops in particular. Extensive web site provides information, lists of members and shops, details of campaigns, news and discussion. A list of UK shops gives addresses and contact details.
Email: eunews@worldonline.nl
Address: Catharijnesingel 82, 3511 GP Utrecht, Netherlands
Tel: +31 30 230 08 20
Fax: +31 30 230 04 40

Network Organic
www.networkorganic.com

Specialities: Lifestyle; directories
Website: New organic-living lifestyle website. It provides on-line directories of growers, suppliers and restaurants. There are also section headings on organic chef and lifestyle.

New Economics Foundation
www.neweconomics.org

Specialities: Organisation; finance; economics
Website: Independent think tank which promotes practical and creative ideas for a fair and sustainable economy. Provides research into durability of products and how to extend product life in order to minimise waste. Website describes the aims of the organisation and provides information on a variety of topics such as alternative currencies, social investment, indicators to sustainability and social accounting and auditing.
Email: info@neweconomics.org
Address: Cinnamon House, 6 - 8 Cole Street, London SE1 4YH
Tel: 020 7407 7447
Fax: 020 7407 6473

New Farm Organics
www.newfarmorganics.co.uk

Specialities: Meat; vegetables; cereals; pulses; wholesaler
Website: Supplies to the organic wholesale market and to various box schemes. Website gives general background information to the
Contact: Jane Edwards
Email: newfarmorganics@zoom.co.uk
Address: J F & J Edwards & Sons, Soulby Lane, Wrangle, Boston, Lincolnshire PE22 2BT
Tel: 01205 870500
Fax: 01205 871001

New Zealand Natural Food Company
www.comvita.com

Specialities: Honey; apitherapy
Website: Supplies some of the finest New Zealand organic honeys and the Comvita range of bee products. Web pages give information on apitherapy, the use of bee products for the treatment of human conditions. The product range is shown on the site. There is a feedback email facility.
Contact: Melina Merryn
Email: info@nznf.co.uk
Address: Unit 3, 55 - 57 Park Royal Road, London NW10 7LP
Tel: 020 8961 4410
Fax: 020 8961 9420

NHR Organic Oils
www.nhr.kz

Specialities: Toiletries; bodycare; aromatherapy products; chocolates; books
Website: The purest organic aromatherapy oils at affordable prices. All oils are presented in clear bottles enabling their vibrant colours to be appreciated. A vast range of organic products from organic lavender shampoo to organic gold, frankincense and myrrh chocolate. Organic essential oils, French certified, available by mail order. The website is very extensive and includes on-line shopping and information pages.
Links: Yes
On-line ordering: Yes
Credit cards: Most major credit cards.
Contact: K Zinovieff & R Manjari
Email: organic@nhr.kz
Address: 10 Bamborough Gardens, London W12 8QN
Tel: 020 8746 0890
Fax: 020 8743 9485

Nonu International
www.sallywigmore.co.uk

Specialities: Noni juice; food supplements; herbal remedies
Website: Pure Noni juice, flavoured Noni juice, Noni tablets, chewable tablets and Noni teabags. Noni fruit contains naturally occurring vitamins, minerals, trace elements, enzymes, beneficial alkaloids, co-factors and plant sterols. The leaves also contain the full spectrum of amino acids which makes these products an excellent source of protein.
Contact: Sally Wigmore
Email: nonu@sallywigmore.co.uk
Address: Sally Wigmore, 5 Queen Square, Lancaster, Lancashire LA1 1RN
Tel: 0800 195 8369
Fax: 07092 296479

Norfolk Organic Gardeners
www.norfolkorganic.mcmail.com

Specialities: Organisation; gardening
Website: Norfolk group of Soil Association and HDRA with aims to promote public awareness in Norfolk of organic methods of farming and gardening. Website gives details of talks, events and visits and provides information on joining. There are links to other regional organic garden societies.
Links: Yes
Contact: Janet Bearman
Email: norfolkorganic@cwcom.net
Address: 6 Old Grove Court, Norwich, Norfolk NR3 3NL
Tel: 01603 403415

Norgrow International Ltd
www.norgrow.com

Speciality: Food industry supplier
Website: Supplies the industry with quality products licensed by the Soil Association: fruit and vegetables, peas, beans & pulses, herbs and spices, essential oils, seeds, grains and lentils, nuts, rices and sugars. Undertakes project partnering, management, supply chain auditing and certification. Website provides comprehensive information about the company and lists in detail both organic and conventional food products available.
Contact: Henri Rosenthal
Email: sales@norgrow.com
Address: Grange Farm Lodge, Leverington Common, Wisbech, Cambridgeshire PE13 5JG
Tel: 01945 410810
Fax: 01945 410850

North East Organic Growers
www.made-in-northumberland.co.uk

Specialities: Vegetables; fruit; herbs; box scheme
Website: A worker's co-operative based at Earth Balance, a sustainable development of farmland, fishing lake, nature reserve and market garden. Website provides general information.
Contact: Bryony Stimpson
Email: neog@care4free.net
Address: West Sleekburn Farm, Bomarsund, Bedlington, Northumberland NE22 7AD
Tel: 01670 821070
Fax: 01670 821026

Northfield Farm Ltd
www.northfieldfarm.com

Specialities: Meat; poultry; turkeys; game
Website: Beef from naturally reared cattle as well as pork, lamb and poultry. Beef comes from a herd of Dexter cattle. On the website prices are given by weight for all cuts of meat available and orders are then placed by telephone.
Contact: Jan McCourt
Email: nthfield1@aol.com
Address: Northfield Farm, Whissendine Lane, Cold Overton, Oakham, Rutland LE15 7ER
Tel: 01664 474271

Nu2rition Ltd
www.nu2rition.com

Speciality: Bodycare
Website: Natural and organic personal care products including hand-made soaps and non-chemical deodorants. On-line shopping is under development and website was under construction at the time of writing.
Credit cards: Visa; Mastercard
Contact: Louise Wood
Email: nu2@nu2rition.com
Address: Lewin House, Dorking Road, Tadworth, Surrey KT20 5SA
Tel: 01737 812205
Fax: 01737 812658

The Nutri Centre
www.nutricentre.com

Specialities: Complementary medicine; resource; books
Website: Europe's leading centre for complementary medicine which houses not only some of the UK's most eminent practitioners but also Europe's leading dispensary and an extensive library and bookshop on complementary medicine, thus integrating all the services under one roof. There are over 22,000 products on offer through the dispensary. The centre will endeavour to answer by email questions on products or medical questions. There is also information on the Education Resource Centre.
Email: enq@nutricentre.com
Address: 7 Park Crescent, London W1N 3HE
Tel: 020 7436 5122
Fax: 020 7436 5171

Nutricia
www.numico.com

Specialities: Food supplements; vitamins; babycare; sports nutrition
Website: Nutricia is part of Royal Numico, global specialists in all aspects of nutrition, ranging from products for babies and infants, lifestyle vitamins and supplements, sports nutrition and specialised clinical products. Website is in English and Dutch and gives company profile, history, investor relations, food safety and quality, and worldwide addresses.
Contact: Chris Blinco
Address: Nutricia Ltd, White Horse Business Park, Trowbridge, Wiltshire BA14 0XQ
Tel: 01225 768381
Fax: 01225 768847

Nutrisport
www.nutrisport.co.uk

Specialities: Food supplements; sports nutrition
Website: Manufacturer of protein, energy and weight-gain mixes. Food supplements for athletes and body builders. Website is divided into proteins and aminos, weight gainers, creatine and energy, vitamins and minerals. On-line fully secure ordering facility.
On-line ordering: Yes
Credit cards: Most major credit cards
Contact: Kevin Hayes
Email: sales@nutrisport.co.uk
Address: Great George Street, Wigan, Lancashire WN3 4DL
Tel: 01942 519111
Fax: 01942 519222

Nutshell Natural Paints
www.nutshellpaints.com

Speciality: Paint
Website: Paints are produced from traditional recipes using natural raw materials. No products are tested on animals. The website provides full details of the product range with a colour chooser facility including paint strengths. There are decorating ideas and recipes. Orders may be placed on-line, by email or by post.
Links: Yes
On-line ordering: Yes
Credit cards: Most major credit cards
Email: info@nutshellpaints.com
Address: PO Box 72, South Brent, Devon TQ10 9YR
Tel: 01364 738 01
Fax: 01364 730 68

Oceans of Goodness
www.seagreens.com

Specialities: Seaweed products; food supplements
Website: Food products manufactured from pure wild seaweeds and marketed under the name Seagreens. Customers are invited to contact Seagreens for nearest stockists of products listed on the site. Website gives background details to the product range.
Email: oceans@probono.org.uk
Address: 1 The Warren, Handcross, West Sussex RH17 6DX
Tel: 01444 400403
Fax: 01444 400493

Oerleman's Foods UK Ltd
www.oerlemans.co.uk

Specialities: Frozen vegetables; frozen fruit
Website: Imports organic frozen vegetables, fruit and oven chips from Germany and Holland and supplies in retail and industrial packaging. Website is available in English and Dutch. It gives details of products available in the organic range and a description of the group, its structure, its production facilities and its plans for the future.
Contact: John Burnett
Email: info@oerlemans.co.uk
Address: The Old Granary, Manor Farm, Aylesby, Grimsby, Lincolnshire DN37 7AW
Tel: 01472 750115
Fax: 01472 877848

Old Macdonalds
www.eio.co.uk

Speciality: Meat
Website: Fresh, Scottish, bio-dynamically-raised organic meat. Information page on organic meat and on-line ordering facility.
Links: Yes
On-line ordering: Yes
Email: info@eio.co.uk
Address: Old Macdonalds Evergreen Internet Organics, No 3, 23 Royal Crescent, London W11 4SN
Tel: 07980 820191

Old Pines Restaurant with Rooms
www.oldpines.co.uk

Specialities: Restaurant; bed & breakfast; holidays
Website: The award-winning restaurant makes use of local products much of it home made, picked and grown. The site gives details of accommodation and tariffs as well as information on the local area.
Contact: Bill & Sukie Barker
Email: goodfood@oldpines.co.uk
Address: Spean Bridge, by Fort William, Highland PH34 4EG, Scotland
Tel: 01397 712324

On The Eighth Day Co-operative Limited
www.eighth-day.co.uk

Specialities: Groceries; Japanese foods; macrobiotic foods; vegetarian & vegan; herbal remedies: café
Website: Vegetarian wholefood shop with almost 1,000 organic lines and café. An extensive website which gives details of the co-operative, its history and ethics. The shop index gives a comprehensive product listing with prices. Mail order is available via email with secure on-line payment. There is also a section on the café with recipes, cookbook and gourmet evenings.
Links: Yes
Credit cards: Most major cards
Contact: Tim Gausden
Email: mail@eighth-day.co.uk
Address: 111 Oxford Road, Allsaints, Manchester M1 7DU
Tel: 0161 273 4878
Fax: 0161 273 4878

Only Natural Products Ltd
www.onlynaturalproducts.co.uk

Specialities: Herbal teas; fruit teas
Website: Producers of herbal and fruit teas. All flavours in the fruit teas are entirely natural. The ingredients of each category of tea are provided on the website. Herbs are described in detail with their history. There is a list of UK stockists. There will shortly be an on-line shopping facility.
Contact: Keith Garden
Email: keithgarden@hotmail.com
Address: Chennels House, Rusper Road, Horsham, West Sussex RH12 5QW
Tel: 01403 251972
Fax: 01403 251972

Opal London
www.opal-london.com

Specialities: Bodycare; aromatherapy products
Website: Leading manufacturer and distributor in the UK of body maintenance products including organic loofahs, massagers, wet and dry body brushes, lotions, gels and oils. Sophisticated website divides shopping zones into seasonal stuff, spa, brushes, lotions and potions, suave massage and style. There is an on-line shopping facility.
On-line ordering: Yes
Credit cards: Most major credit cards
Contact: Nina Hobart
Email: sales@opal-london.com
Address: 65 Waterloo Road, London NW2 7TS
Tel: 020 8450 7834
Fax: 020 8450 3293

Optima Healthcare Ltd
www.optimahealthcare.co.uk

Specialities: Herbal remedies; food supplements; vitamins; bodycare; aloe vera
Website: Leading manufacturer and distributor of aloe vera products with tablets, skin gels, juice drinks, digestive and colon cleanse aids, toothpaste and mouthwash products and a raw aloe range of toiletries and allergenic skin care products. Products are listed alphabetically on the website with on-line shopping facility. A Find-A-Cure facility was being redesigned at the time of writing. There is a list of major stockists but also an email facility for locating the nearest independent healthfood retailer stockist.
On-line ordering: Yes
Credit cards: Most major cards
Contact: Hadrian Gower
Email: info@optimahealthcare.co.uk
Address: 47 - 48 St Mary Street, Cardiff, South Glamorgan CF10 1AD, Wales
Tel: 02920 388422
Fax: 02920 233010

Ord River Tea Tree Oil Pty Ltd
www.ordriver.co.uk

Specialities: Tea tree oil; aromatherapy products; essential oils
Website: The world's first complete range of natural tea tree products free of artificial preservatives, fragrances and colours. The products are ideal for those with sensitive skin or who suffer from skin allergy complaints. Website details benefits and research background of tea tree oil. There is an on-line catalogue with ordering facility.
On-line ordering: Yes
Credit cards: Most major credit cards
Contact: David Chapman
Email: info@ordriver.co.uk
Address: The Cedar Suite, Winchester Road, Alresford, Hampshire SO24 9EZ
Tel: 01962 734080
Fax: 01962 733289

The Organic Advisory Service
www.efrc.com

Specialities: Organisation; advice; farming; agriculture
Website: The UK's leading organic agricultural research and development centre and educational charity. Provides education, advice and research on organic food and farming. A soil analysis service is available. The website provides information on research, publications, education, news and policy, student packs and community. There are also details of the organic demonstration farm network.
Links: Yes.
Email: elmfarm@efrc.com
Address: Elm Farm Research Centre, Hamstead Marshall, Newbury, Berkshire RG20 0HR
Tel: 01488 658298
Fax: 01488 658503

Organic and Natural Foods
www.organic-and-natural.co.uk

Specialities: Home delivery; fruit; vegetables; meat; wines; box scheme
Website: Independent home delivery company retailing organic fruit and vegetables, meats, fish, bread and dairy products alongside a comprehensive range of organic and natural wholefoods and eco-friendly products. National service with one-stop shopping facility for organic produce available through the website. Where appropriate items can be ordered by weight.
Links: Yes
On-line ordering: Yes
Email:
nanas@organicandnat.demon.co.uk
Address: Unit 3, Atlas Transport Estate, Bridges Wharf, Bridges Court, York Road, Wandsworth, London SW11 3RE
Tel: 020 7924 7800
Fax: 020 7294 7900

The Organic Baby Food Company
www.tetbury.com/baby

Speciality: Baby food
Website: Provides 100% organic fruit and vegetable purées, frozen in handy ice cube form. The method of manufacture ensures that the colour, flavour and texture is retained so that babies get the real taste of fruit and vegetables. Website provides a price list and listing of products with baby age recommendations. Orders may be placed for home delivery.
Contact: Libby Townsend
Email: libbytownsend@cs.com
Address: The Stables, Manor Farm, Chavenage, Tetbury, Gloucestershire GL8 8XW
Tel: 01666 505616
Fax: 01666 504616

The Organic Beef Co
www.theswaninn-organics.co.uk

Specialities: Meat; farm shop; hotel; restaurant
Website: Organic farm shop offering a selection of 1,000 plus organic products including organic beef produced on their Inkpen farm. The restaurant menu features organic beef dishes and organic desserts. Mail order is under consideration and potential customers should contact direct. Website gives shop, hotel and restaurant details as well as directions.
Contact: Bernard Harris
Email:
enquiries@theswanninn-organics.co.uk
Address: The Old Craven Arms, Inkpen, Hungerford, Berkshire RG17 9DY
Tel: 01488 668326
Fax: 01488 668306

Organic Centre for Wales
www.organic.aber.ac.uk

Specialities: Organisation; courses; farming; agriculture
Website: Runs degree courses on organic agriculture and gives advice and conducts research. Useful statistics on organic farming are available on the website. Website topics include thinking of going organic, research, training and education, demonstration farm network, advisory services, publications and document library.
Email: organic-helpline@aber.ac.uk
Address: University of Wales, Aberystwyth, Ceredigion SY23 3AL, Wales
Tel: 01970 622248
Fax: 01970 622238

Organic Concentrates
www.6-x.co.uk

Specialities: Fertilisers; gardening
Website: Producers of organic fertilisers under the brand '6X' which are sold through garden centres as well as by mail order. Website provides details of stockists as well as a link for on-line ordering through Gardendirect. Mail order prices for phoned or faxed orders are provided.
Links: Yes
Credit cards: Most major cards
Contact: Chris Green
Email: Organic6x@6-x.co.uk
Address: 3 Broadway Court, Chesham, Buckinghamshire HP5 1EN
Tel: 01494 792229
Fax: 01494 792199

Organic Connections International
www.organic-connections.co.uk

Specialities: Fruit; vegetables; box scheme; home delivery; pet food
Website: Company deals solely with organic produce both as a grower, packer and importer. The website gives details of some of their growers. There is an order form for boxes and other produce on the website which can be submitted electronically. Orders can also be telephoned for personal service.
Credit cards: Visa; Mastercard, Switch; Amex; Delta
Contact: Karen & Edwin Broad
Email: sales@organic-connections.co.uk
Address: Riverdale, T
own Street, Upwell, Wisbech, Cambridgeshire PE14 9AF
Tel: 01945 773 374
Fax: 01945 773 033

The Organic Consultancy
www.organic-consultancy.com

Speciality: Consultancy
Website: Helps companies develop organic processed food and drink. The website explains the basic principles of organic food manufacture, summarises the activities of the consultancy, and enables contact to be made to discuss requirements. There are articles and a client list.
Contact: Simon Wright
Email: simon@organic-consultancy.com
Address: 101 Elsenham Street, London SW18 5NY
Tel: 020 8870 5383
Fax: 020 8870 8140

Organic Days
www.organicdays.co.uk

Specialities: Chocolate; sweets; jams; teas; dried products
Website: Whenever possible certified organic products are used. There are special products for cocoa-free, cow's milk protein-free and gluten-free. The intention is to offer a wide range of organic products that can be ordered in the simplest possible way over the internet. Organic chocolates, sweets, preserves, teas, pulses, dried fruits, cereals, nuts and other organic products are listed on the website for on-line ordering.
On-line ordering: Yes
Credit cards: Most major cards
Contact: Adrian Lauchlan
Email: info@organicdays.co.uk
Address: Unit 11B,
Crusader Industrial Estate,
167 Hermitage Road,
London N4 1LZ
Tel: 020 8802 1088
Fax: 020 8802 3862

The Organic Delivery Company
www.organicdelivery.co.uk

Specialities: Groceries; box scheme; fruit; vegetables; wines; vegetarian
Website: London organic box scheme. The food is grown in Devon. There is secure on-line ordering. Also supplies kitchen equipment and clothing.
Links: Yes
On-line ordering: Yes
Contact: John Barrow
Email: info@organic-delivery.co.uk
Address: 7 Willow Street,
London EC2
Tel: 020 7739 8181
Fax: 020 7613 5800

The Organic Farm Shop
www.theorganicfarmshop.co.uk

Specialities: Meat; vegetables; farm shop; groceries
Website: Organic fresh vegetables, meat, dairy products and general groceries. The website gives a listing of produce for sale and orders can be faxed, phoned or emailed. There are information pages on the farm, the shop, the café, matters organic, the garden and educational visits. There is also an extensive newspage.
Credit cards: Most major cards
Email: info@theorganicfarmshop.co.uk
Address: Abbey Home Farm,
Burford Road, Cirencester,
Gloucestershire GL7 5HF
Tel: 01285 640441
Fax: 01285 644827

Organic Feed Company
www.allenandpage.com

Speciality: Animal feed
Website: The company deals in specialist animal feeds with no synthetic ingredients such as artificially manufactured vitamins and amino acids. Website provides details of organic standards and ingredients. There is information on organic feeds for poultry, pigs, cattle, sheep and goats. There is a listing of organic feed stockists which is searchable by county.
Email: organic@allenandpage.co.uk
Address: Allen & Page,
Shipdham, Thetford,
Norfolk IP25 7SD
Tel: 01362 822 903
Fax: 01362 822 910

Organic Food
www.organicfood.co.uk

Specialities: Information; resource; directory
Website: On-line organic magazine format website which includes a local grower's directory. Website provides top 10 reasons to go organic, a database of more than 800 organic food outlets, reviews of home delivery services, chat zones, articles and a herb library
Email: info@organicfood.co.uk

The Organic Food Federation
www.organicfood.co.uk /off/index.html

Specialities: Organisation; farming; agriculture; gardening
Website: Acts as a focal point for members to establish contact with one another, communicates information, ensures high standards, and represents members with government control bodies. Also provides a certification and inspection service. Website gives information on complying with organic principles and the benefits of organic farming as well as guidance notes.
Email: organicfood@freenet.co.uk
Address:
1 Mowles Manor Enterprise Centre, Etling Green, East Dereham, Norfolk NR20 3EZ
Tel: 01760 720444

Organic Gardening Catalogue
www.organiccatalog.com

Specialities: Seeds; vegetable seeds; wildflower seeds; vegan gardening; gardening
Website: Official catalogue of the HDRA. On-line ordering for mail order catalogue supplying seeds and other related items. The catalogue also sells many products which may be used in the veganic garden. Non-organic seeds are offered where no organic variety is yet available,
Links: Yes
On-line ordering: Yes
Credit cards: Most major cards
Catalogue: Yes (gratis)
Contact: Mike Hedges
Email: chaseorg@aol.com
Address: River Dene Estate, Molesey Road, Hersham, Surrey KT12 4RG
Tel: 01932 253666
Fax: 01932 252707

The Organic Herb Trading Company
www.organicherbtrading.com

Specialities: Wholesaler; herbs; essential oils
Website: Two decades in the business of growing, processing, importing and distribution of organic herbs. The company currently handles nearly 1,000 plant and pure plant derived products including dried herbs and spices, essential oils, pressed oils, macerated oils, tinctures, extracts, flower waters and flavours for use in beverages, herbal medicines and cosmetics. Website gives general details of sourcing, products and services.
Contact: Mike Brook
Email: info@organicherbtrading.com
Address: Court Farm, Milverton, Somerset TA4 1NF
Tel: 01823 401205
Fax: 01823 401001

Organic Holidays
www.organicholidays.co.uk

Speciality: Holidays
Website: An international on-line guide to bed and breakfast, small hotels and guest houses where organic produce is used according to availability. It also includes accommodation to rent on organic farms and organic smallholdings.
Email: lindamoss@organicholidays.co.uk
Address: Tranfield House, 4 Tranfield Gardens, Guiseley, Leeds, Yorkshire LS20 8PZ
Tel: 01943 871 468
Fax: 01943 871 468

Organic India Ltd
www.organicindia.co.uk

Specialities: Indian food; ready meals
Website: The first company in the UK to produce organic Indian ready meals. The ingredients used are all natural. Meat and poultry is sourced from small organic farms. Website describes the ready meals in the range and lists their ingredients in detail. There is a listing of outlets with contact details where the foods may be purchased.
Links: Yes
Email: sales@organicindia.co.uk
Address: 30 Queens Court, Kenton Lane, Kenton, HA3 8RL
Tel: 020 8907 3727

Organic Kosher Foods
www.organickosher.co.uk

Specialities: Kosher foods; halal; meat; poultry; fish; box scheme
Website: Kosher, halal and conventional foods all organically sourced. There is a box scheme for home delivery. Website provides a list of outlets which stock some of their products. Requirements for box scheme can be filled in on-line and the company will then contact regarding credit card payment and delivery. They can also supply on a wholesale basis.
Links: Yes
Credit cards: Most major cards
Contact: Leon Pein
Email: sales@organickosher.co.uk
Address: OK Foods, PO Box 3079, Barnet, Hertfordshire EN5 4ZD
Tel: 0800 458 567437
Fax: 0800 980 6198

Organic Living
www.organicliving.co.uk

Speciality: Magazine
Website: Consumer title with regular articles on health, children, food and wine, fashion and recipes. The website provides general information including display advertising rates.
Contact: Nicky Keane
Email: info@organicliving.co.uk
Address: 4 South Park Road, Harrogate, North Yorkshire HG1 5QU
Tel: 01423 705052
Fax: 01423 705051

The Organic Marketplace
www.organicmarketplace.co.uk

Speciality: Directory
Website: On-line directory of UK organic businesses and products. Search by product or service category or by area. It covers organic buying and selling and has an on-line selling facility. There are pages of news, events and links. Organic events and news items can be posted free on the site.
Links: Yes
Email:
enquiries@organicmarketplace.co.uk
Tel: 01452 864 481

Organic Matters
www.organicmatters.co.uk

Speciality: Milk
Website: Group of 6 organic dairy farmers producing organic milk. Website gives background information and details of different types of milk available. It also gives details of the individual farmers.
Contact: D Stacey
Email: sales@organicmatters.co.uk
Address: Leigh Manor, Ministerley, Shropshire SY5 0EX
Tel: 01743 891929

Organic Meats & Producers Scotland Ltd
www.jamesfieldfarm.co.uk

Specialities: Meat; poultry; game
Website: Next day home delivery. The on-line shop offers organic beef, lamb and poultry as well as additive-free free-range pork. There is also a printable

order form. There is a picture gallery and a brief history of the farm.
On-line ordering: Yes
Credit cards: Most major cards
Contact: Ian Miller
Email: jamesfieldfarm@excite.co.uk
Address: Jamesfield Farm, Abernethy, Fife KY1 6EW, Scotland
Tel: 01738 850498
Fax: 01738 850741

Organico
www.organico.co.uk

Specialities: Baby foods; juices; olive oil; ready meals; chocolate
Website: Producer of gourmet organic baby foods and formulas. Mail order is available. Brand names include Radici, Vitalia and Babynat. Site provides illustrated descriptive survey of their products. There is also an extensive and informative page answering the most common questions about organic foods.
Contact: C Redfern
Email: info@organico.co.uk
Address: 63 High Street, London N8 7QB
Tel: 0281 340 0401
Fax: 0281 340 0402

Organic Oxygen
www.organic-oxygen.co.uk

Specialities: Wines; hampers; seeds; vegan; bodycare; gifts
Website: Offers a range of organic, GMO and gluten-free foods, wines, beers, champagne, cider and also seeds. Big selection of hampers and different sized boxes of fruit, vegetables and cheese with free delivery in England and Wales. Also features vegan shoes, sarongs and batiks.
Links: Yes
On-line ordering: Yes
Credit cards: Most major credit cards
Email: help@organic-oxygen.co.uk
Address: 89a Manor Road, Banbury, Oxfordshire OX16 7JL
Tel: 01295 252 496
Fax: 0870 121 7520

Organic Resource Agency Ltd
www.growscompost.co.uk

Specialities: Consultancy group; composting
Website: Specialises in the green recycling of organic waste from supermarkets. An innovative and unique project in which Waitrose and Sainsbury's are working together with Sheepdrove Organic Farm to find an environmentally friendly way of disposing of fruit, vegetable and flower waste from their stores. There is a question and answer page, press releases and general information.
Email: ora@efrc.com
Address: Elm Farm Research Centre, Hamstead Marshall, Newbury, Berkshire RG20 0HR
Tel: 01488 657 658
Fax: 01488 658 503

Organic Roots
www.organicroots.co.uk

Speciality: Farm shop
Website: Farm shop close to Birmingham dedicated to the supply of a wide range of organic products. Items can be selected from the indexed product lists on the website and then ordered by phone, fax or email. There is a recipe page as well as an FAQ page.
Credit cards: All major credit cards
Contact: Bill Dinenage
Email: info@organicroots.co.uk
Address: Crabtree Farm, Dark Lane, Kings Norton, Birmingham, West Midlands B38 0BS
Tel: 01564 822294
Fax: 01564 829212

The Organic Shop (Online) Ltd
www.theorganicshop.co.uk

Specialities: Vegetables; meat; home delivery; box scheme; baby food; groceries
Website: Committed to offer certified organic food and environmentally friendly products, to reduce waste and environmental pollution and to pay and charge a fair price. Product range includes organic boxes, fruit and vegetables, meat, dairy, grocery, wine, beer and spirits, baby food and household products. A record of past orders is stored on the website.
On-line ordering: Yes
Credit cards: Most major credit cards
Catalogue: Brochure can be downloaded or emailed
Email: info@theorganicshop.co.uk
Address: Central Chambers, London Road, Alderley Edge, Cheshire SK9 7YG
Tel: 0800 195 7844
Fax: 0870 121 7520

Organic Spirits Company
www.junipergreen.org

Speciality: Spirits
Website: Juniper Green is the world's first organic London Dry gin. All the ingredients are 100% organic. Also organic vodka, organic white rum and organic spiced rum. The website provides historical information, tasting suggestions, information on organics and international contact addresses.
Email: office@londonandscottish.co.uk
Address: Meadow View House, Tannery Lane, Bramley, Surrey GU5 0AB
Tel: 01483 894 650
Fax: 01483 894651

Organics To Go
www.organics2go.co.uk

Specialities: Box scheme; home delivery; vegetarian & vegan
Website: Co-operative business set up to support organic growers to get their fresh produce directly to customers. Late evening and early morning deliveries in Cardiff, Carmathenshire, Bristol, East London and Hampstead. Box scheme allows customer to choose items each week. On-line order form for existing customers. There is a link to www.veganvillage.co.uk.
Links: Yes
Contact: Roger Hallam
Email: mail@organics2go.co.uk
Address: Werndolau Farm, Golden Grove, Carmarthen, Carmarthenshire SA32 8NE, Wales
Tel: 0800 458 2524
Fax: 01558 668088

Organictrade Ltd
www.organictrade.co.uk

Specialities: Wholesalers; nuts; dried fruits; pulses; cereals
Website: Large scale importer of dried goods, pulses and cereals for bulk supply by the 1/4, 1/2 and 1 ton. Website provides order form, prices, and product descriptions. There is a page for terms and conditions.
Contact: Craig Whitelaw
Email: info@organictrade.co.uk
Address: Premier House, 325 Streatham High Street, London SW16 3NT
Tel: 020 8679 8226
Fax: 020 8679 8823

Organic Trail
www.organictrail.freeserve.co.uk

Speciality: Box scheme
Website: Box scheme delivering to the local area with next-day nationwide delivery service. Freshly harvested produce comes from local farms. Box contents cannot be customised. Weekly or fortnightly service.
Contact: Jim Lawlor
Email: jim@organictrail.freeserve.co.uk
Address: 10 St Paul's Court, Stony Stratford, Milton Keynes, Buckinghamshire MK11 1LJ
Tel: 01908 568952
Fax: 01908 568952

The Organic Wholesale Company
www.organicwholesale.co.uk

Speciality: Wholesalers
Website: Wholesalers of high quality fresh organic produce at sensible prices delivered throughout the UK. On the website there is an email form to request a full product list.
Email: trade@organicwholesale.co.uk
Address:
8 - 13 Mahatma Ghandi Industrial Estate, Milkwood Road, London SE24 0FJ
Tel: 020 7798 8988
Fax: 020 7737 7785

The Organic Wine Company
www.organicwinecompany.com

Specialities: Wines; spirits; juices
Website: The website provides an explanation of organic wine. The company supplies individuals as well as the trade and specialises in vegetarian restaurants, healthfood shops and delis. The wine list is available on request by email, phone or letter. Day trips to collect organic wine duty free from France are organised on a regular basis.
Credit cards: Visa; Mastercard
Catalogue: Wine list
Email: afm@lineone.net
Address: PO Box 81, High Wycombe, Buckinghamshire HP13 5QN
Tel: 01494 446557
Fax: 01494 446557

The Organic Wool Company
www.organicwool.co.uk

Specialities: Wool; babycare; textiles
Website: Baby blankets, rugs, throws, scarves and shawls manufactured in Wales from organic wool, natural wool and rare breed wool which are processed without bleach or harmful chemical finishes. An order form can be downloaded and there is a mail order service. There is also information on organic sheep, the wool, organic textile standards and selling organic wool.
Catalogue: Brochure
Email:
douglaswhitelaw@organicwool.co.uk
Address: Jasmine House, School Road, Barnack, Stamford
Cambridgeshire PE9 3D7
Tel: 01780 740021

Organics Direct
www.organicsdirect.co.uk

Specialities: Fruit; vegetables; juicers; box schemes; wines; baby products
Website: There is a useful coding system for people on special diets. Company supplies vegetarian organic groceries and is totally committed to organic produce and to campaigning against GM food. Delivers to anywhere on the UK mainland. All produce comes from small farmers and growers. Considerable variety of non-food lines from clothing to kitchen equipment to nappies. There is a recipe and hints page. Prices are up-dated weekly.
On-line ordering: Yes
Catalogue: Brochure
Contact: John Barrow
Email: info@organicsdirect.co.uk
Address: 7 Willow Street,
London EC2
Tel: 020 7729 2828
Fax: 020 7613 5800

Organics-on-Line
www.organics-on-line.com

Specialities: Internet trading site; business to business
Website: Facilitates trade direct from producer to buyer by targeting business to business trade in organics creating links between buyers and producers. Website gives news, suppliers, buyers and sponsors as well as a helpdesk. Suppliers and buyers can register their products and interests on-line.
Contact: Annette Ward
Email: info@organics-on-line.com
Address: Park House, Gunthorpe Hall, Melton Castle, Norfolk NR24 2PA
Tel: 01263 861991
Fax: 01263 860964

OrganicSeeds.co.uk
www.organicseeds.co.uk

Specialities: Seeds; vegetable seeds; herb seeds; soft fruit gardening
Website: Full on-line catalogue indexed by varieties. Printable order form. A companion planting chart may also be ordered.
Links: None
Email: sales@organicseeds.co.uk
Address: PO Box 398, Ipswich, Suffolk IP9 2HU
Tel: 01473 310118

Organic-research.com
www.organic-research.com

Specialities: Resource; farming; agriculture
Website: Resource for anyone studying any aspect of organic farming, food and related issues. Links to research papers and projects as well as details of international conferences and educational courses. There are a number of free visitor areas on the site but the premier content is only available by membership. There is a searchable database of over 100,000 informative abstracts of articles from research journals, conference proceedings, reports and books. There are news headlines, a selection of research papers and a list of educational courses by region among the many other site headings. There is a free 30 day trial.
Links: Yes
Email: cabi@cabi.org
Address: CAB International, Wallingford, Oxfordshire OX10 8DE
Tel: 01491 832111
Fax: 01491 833508

Organigo
www.organigo.co.uk

Specialities: Retail shop; vegetarian & vegan
Website: Retail shop specialising in vegetarian and vegan produce. Bread, fruit and vegetables, dairy products, ice cream, wines, beers, ciders and spirits. Special Welsh section. Website under construction at time of writing.
Contact: James Goodsir
Email: shop@organigo.co.uk
Address: 1 Tower Street, Crickhowell, Powys NP8 1PL, Wales
Tel: 01873 811112
Fax: 01873 811037

The Original Seagrass Company Ltd
www.original-seagrass.co.uk

Speciality: Flooring
Website: Natural floor coverings supplied direct without intermediary mark-ups. Prices include complete installation. Website illustrates and lists product range with prices. There is an email contact form for applying for an estimate.
Contact: David & Janet Green
Email: original.seagrass@virgin.net
Address: Shrewsbury Road, Craven Arms, Shropshire SY7 9NW
Tel: 01588 673 666
Fax: 01588 673 667

Origins
www.origins.com

Specialities: Bodycare; skincare
Website: The website is from the parent company in the US and on-line ordering is not UK oriented. There is, however, a store locator for the UK and the on-line catalogue does detail their product range.
On-line ordering: Yes
Email: originsonlinetech@origins.com
Address: 51 King's Road, London SW3
Tel: 020 7833 6715

Oscar Mayer Ltd
www.oscarmayer.co.uk

Speciality: Ready meals
Website: Prepares ready meals for multiple retailers and supermarkets. Site

gives general details of the company but was under construction at time of writing.
Contact: B J Rodgers
Email: webmaster@oscarmayer.co.uk
Address: Furnham Road, Chard, Somerset TA20 1AA
Tel: 01460 63781
Fax: 01460 67847

Otley College
www.otleycollege.ac.uk

Specialities: Courses; farming; agriculture; gardening
Website: Otley provides an extensive range of courses in organics, conservation and sustainable systems. These include commercial organic production, holistic horticulture and horticultural therapy. The website provides details of the various courses available together with contact details. There are also indexes for further and higher education.
Email: course_enquiries@otleycollege.ac.uk
Address: Otley, Ipswich, Suffolk IP6 9EY
Tel: 01473 785543
Fax: 01473 785553

Out Of This World
www.ootw.co.uk

Speciality: Supermarket
Website: This small chain of ethical and organic supermarkets carries over 4,000 products. It is a consumer co-op with some 18,000 members. There is a mail order catalogue listing organic food, drinks, delicatessen, natural healthcare, eco-friendly house and garden products and fairtrade crafts. Website includes pages on healthy eating, environmental sustainability, fairtrade, animal welfare and community development. It gives details of its stores in Cheltenham, Newcastle and Nottingham. Membership information and how to join the co-operative is provided.
Contact: Brian Hutchins
Email: info@ootw.co.uk
Address: 106 High Street, Gosforth, Newcastle upon Tyne, Tyne and Wear NE3 1HB
Tel: 0191 213 5377
Fax: 0191 213 5378

Outside In (Cambridge) Ltd
www.outsidein.co.uk

Specialities: Seasonal affective disorder; light therapy
Website: A leader in light therapy for treating SAD and other body clock related problems. The website contains information on seasonal affective disorder, sleep, jetlag and alzheimer's. There are details of the various products available to combat these problems and on-line ordering. There are also details on the home-trial system which allows customers to ensure that the products are what they actually need.
On-line ordering: Yes
Credit cards: Visa; Mastercard; Switch; Delta; Amex
Email: info@outsidein.co.uk
Address: 21 Scotland Road Estate, Dry Drayton, Cambridge, Cambridgeshire CB3 8AT
Tel: 01954 211 955
Fax: 01954 211 956

Oxfam Fairtrade
www.oxfam.org.uk

Specialities: Fairtrade; gifts
Website: Part of the Oxfam general website. Catalogue of food, gifts, clothing jewellery and stationery - fairtrade products where producers have been paid a fair price for their work. Oxfam works with over 160 producers organisations in some 30 countries. Catalogue can be ordered from the website. There are celebrity recipes and many fairtrade articles.
Links: Yes.
On-line ordering: Yes
Email: oxfam@oxfam.org.uk
Address: 30 Murdock Road, Bicester, Oxfordshire OX6 7RF
Tel: 01869 355100

Pallet Display Systems Ltd
www.palletlegs.com

Specialities: Recycling; pallets
Website: Recycling of wooden and plastic pallets to create benches, displays, wheeled trolleys and many other applications with pallet legs and accessories. Website gives an insight into the system and how it works and the company behind it. It illustrates the usefulness and flexibility of the product range and the accessories which can be used in horticulture, industrial, retail, workshops, agriculture, construction, boating, bridging, furniture and garages.
Address: Bentley Lane, Walsall, West Midlands WS2 8TP
Tel: 01922 632575
Fax: 01922 632583

Panacea (Health) UK
www.panaceahealth.co.uk

Specialities: Food supplements; herbal remedies; bodycare
Website: Panacea has introduced many innovative and quality products to the UK healthcare market. They are suppliers of healthfood products and natural toiletries. The website gives details of speciality products as well as products for allergies, eye disorders and stress.
Links: Yes
Contact: Ursula Shrimanker
Email: panacea@btinternet.com
Address: Unit 20, Hallmark Trading Estate, Fourth Way, Wembley, Middlesex HA9 0LB
Tel: 020 8795 3730
Fax: 020 8904 2489

A R Parkin Ltd
www.arparkin.co.uk

Specialities: Spices; food industry supplier
Website: Manufacturers and suppliers of spices, seasonings, ingredients and flavours to the food industry. There is a product listing on the website and information on concept to customer.
Contact: Tony Hilditch
Email: enquiries@arparkin.co.uk
Address: Unit 8, Cleton Street Business Park, Cleton Street, Tipton, West Midlands DY4 7TR
Tel: 0121 557 1150

Pascoes
www.pascoes.co.uk

Speciality: Pet food
Website: Producers of organic cat and dog foods. Dedicated to producing high quality nutritionally sound dog foods without using unnecessary artificial ingredients. Website provides information on the foods and a list of national stockists. There is an email form to enquire about local stockists. Nutritional information on the individual foods is given in detail. Pascoes also run a Breeder's Club with a complete service including home delivery.
Links: Yes
Email: info@pascoes.co.uk
Address: Dunball Wharf, Dunball, Bridgwater, Somerset TA6 4TA
Tel: 01278 444555
Fax: 01278 685251

Pashley Cycles
www.pashley.co.uk

Speciality: Cycles
Website: Hand-built bikes, trikes, recumbents, unicycles, tandems and all kinds of cycles, details of which are available on the website with photos, specifications, sizes and prices. On-line ordering is available.
Links: Yes
On-line ordering: Yes
Email: hello@pashley.co.uk
Address: Masons Road, Stratford Upon Avon, Warwickshire CV37 9NL
Tel: 01789 292 263
Fax: 01789 414 201

Paskins Hotel
www.paskins.co.uk

Specialities: Hotel; restaurant; vegetarian; holidays
Website: Most of the food served is organic and the hotel makes a speciality of its vegetarian breakfasts. The website provides amongst other things, a graphic description of the food offered at breakfast, details of the tariff, a booking email form and a brochure download facility. There is also a virtual tour of Brighton.
Catalogue: Brochure
Contact: R Marlowe
Email: welcome@paskins.co.uk
Address: 18/19 Charlotte Street, Brighton, East Sussex BN2 1AG
Tel: 01273 602476
Fax: 01273 621973

Passion For Life Products Ltd
www.passionforlife.com

Specialities: Health products; natural remedies
Website: Mail order company offering a range of natural (and where possible organic) products from around the world. Categories available include bodycare, health care, women and men's health, oral health, dietary supplements, weight loss, bones and joints, specific ailments, aromatherapy, household products, food and drink.
On-line ordering: Yes
Credit cards: Most major credit cards
Catalogue: Yes
Email: info@passionforlife.com
Address: 21 Heathmans Road, London SW6 4TJ
Tel: 0800 096 1121
Fax: 020 7384 0444

Pause
www.thehoxtoncollective.co.uk

Specialities: Juice bar; café
Website: The organic juice bar is run by the Hoxton Collective and details of their activities are given on the website.
Contact: Ian Hickinbotham
Address: 11 New Inn Yard, London EC2A 3EY
Tel: 020 7729 1341

Pegasus Pushchairs Ltd
www.allterrain.co.uk

Speciality: Pushchairs
Website: All terrain pushchairs that are light, compact and manoeuvrable. Website provide advice on choosing models and illustrates the range and the accessories. There is a regional list of retail outlets in the UK.
Email: info@allterrain.co.uk
Address: Westbridge, Tavistock, Devon PL19 8DE
Tel: 01822 618077

Penrhos Court Hotel and Restaurant
www.penrhos.co.uk

Specialities: Restaurant; hotel
Website: Soil Association certified hotel. There is information on-line on getting married at Penrhos, corporate meetings and training, and eco-organic meetings and seminars.
Contact: Martin Griffiths
Email: info@penrhos.co.uk
Address: Kington, Herefordshire HR5 3LH
Tel: 01544 230720
Fax: 01544 230754

PENTRE BACH HOLIDAY COTTAGES
www.pentrebach.com

Specialities: Cottages; holidays
Website: Holiday cottages in Snowdonia National Park. Free range eggs and wide variety of own grown organic fruit and vegetables. Green practices and permaculture in action. Website gives illustrated details of cottages, availability and prices. There is also regional information.
Contact: Margaret Smyth
Email: orgd@pentrebach.com
Address: Pentre Bach,
Llwyngwril, Nr Dolgellau,
Gwynedd LL37 2JU, Wales
Tel: 01341 250294
Fax: 01341 250885

PERFECTLY HAPPY PEOPLE LTD
www.phpbaby.com

Specialities: Nappies; baby clothes; babycare
Website: Company specialises in Kooshies environmentally-friendly, washable nappies and a wide range of baby items, clothing and equipment. On-line catalogue covers bedroom and bathroom, clothes, nappies and training, kitchen and feeding, toys and playtime, health and safety, swimming, outdoors and travel.
On-line ordering: Yes
Credit cards: Most major cards
Catalogue: Yes
Email: info@phpbaby.com
Address: 31-33 Park Royal Road,
London NW10 7LQ
Tel: 0870 607 0545
Fax: 020 8961 1333

PERMACULTURE ASSOCIATION (BRITAIN)
www.permaculture.org.uk

Specialities: Organisation; permaculture; courses
Website: The national UK organisation promoting permaculture. Provides information, contacts and research. Website gives information on the organisation and its structure, on groups and projects and on courses and educational resources available. There is membership information including a printable membership form.
Email: office@permaculture.org.uk
Address:
BCM Permaculture Association,
London WC1N 3XX
Tel: 07041 390170
Fax: 0113 2621718

PERMACULTURE MAGAZINE & PERMANENT PUBLICATIONS
www.permaculture.co.uk

Specialities: Magazine; books; videos
Website: Publishers of *Permaculture Magazine*, and books on all aspects of sustainable living. Also videos. The website provides a subscription form and details for the magazine. Sample articles are available on-line. There is also a *Permaculture Magazine* information service.
Email: info@permaculture.co.uk
Address: The Sustainability Centre,
East Meon, Hampshire GU32 1HR
Tel: 01730 823311
Fax: 01730 823322

PERO FOODS
www.pero-dogfood.co.uk

Speciality: Pet food
Website: A complete range of organic food for dogs and cats. Carefully selected ingredients blended with vitamins, minerals and trace elements to provide a truly balanced diet. Ingredients are produced without chemical fertilisers, herbicides or pesticides and contain no GMOs.
Contact: Dewi Parry
Email: pero@pero.fsbusiness.co.uk
Address: Llawr Ynis, Betws-y-Coed,
Conwy LL25 0PZ, Wales
Tel: 01690 710457
Fax: 01690 710032

Pershore College
www.pershore.ac.uk

Specialities: Courses; consultancy; gardening; agriculture
Website: Organic fruit and vegetable production unit. Training courses and consultancy. Details of the courses and their requirements are given on the website as well as an application form. There is also information on the Plant Centre, Avonbank Nurseries and Avonbank Fruit.
Contact: John Edgeley
Email: postmaster@pershore.ac.uk
Address: Avonbank, Pershore, Worcestershire WR10 3JP
Tel: 01386 552443
Fax: 01386 556528

Pertwood Organics Co-operative Ltd
www.pertwood-organics.co.uk

Speciality: Box scheme
Website: Farm based box scheme for Wiltshire and the surrounding counties. In addition there is a special order service and a bulk order service. A price list may be had on application.
Links: Yes
Contact: Miranda Tunnicliffe
Email: mail@pertwood-organics.co.uk
Address: The Old Barn at Lords Hill, Lower Pertwood Farm, Longbridge, Deverill, Warminster, Wiltshire BA12 7DY
Tel: 01985 840646
Fax: 01985 840649

Pesticide Action Network
www.pan-uk.org

Specialities: Organisation; pest control
Website: The network works nationally and internationally with like-minded groups and individuals concerned with health, environment and development to eliminate the hazards of pesticides and prevent unnecessary expansion of use and to increase the sustainable and ecological alternatives to chemical pest control. The website provides news updates, publication list, facts and figures, project information and a site search facility.
Links: Yes
Email: admin@pan-uk.org
Address: Eurolink Centre, 49 Effra Road, London SW2 1BZ
Tel: 020 7274 8895
Fax: 020 7274 9084

Pestwatch (Bristol)
www.pestwatch.net

Specialities: Pest control; gardening
Website: Advice and training in sustainable pest management. Website covers pest identification, training, research, gardener's advice, environment services, farming, urban and industrial aspects. There are sections on garden design and tips for gardeners.
Contact: Michael L Warnes
Email: info@pestwatch.net
Address: Alpha Cottage, 8 Merrywood Close, Southville, Bristol BS3 1EA
Tel: 0117 9634 194
Fax: 0973 444862

Pete & Johnny's PLC
www.p-j.co.uk

Speciality: Smoothies
Website: Smoothies made by crushing the plumpest, finest fruits into a thick juicy pulp with nothing added and nothing taken away. Sophisticated web site with graphics.
Contact: Ellen Wilson
Email: jos@p-j.co.uk
Address: 15 Lots Road, Chelsea Wharf, London SW10 0QT
Tel: 020 7352 0276
Fax: 020 7376 7101

Pharmavita
www.pharmavita.co.uk

Specialities: Food supplements; bodycare
Website: International distributor of natural products made to pharmaceutical standards. Leading brands of supplements, creams, lotions and shampoos. Website is designed to help potential customers to understand the products and make an informed decision. Products are listed and there is an electronic order form and price list.
Contact: Nigel Fawkes
Email: admin@pharmavita.co.uk
Address: PO Box 3379 Wandsworth, London SW18 4WZ
Tel: 020 8870 5533
Fax: 020 8877 1111

Phoenix Community Stores
www.phoenixshop.com

Specialities: Retail shop; bakery
Website: Large range of organic foods available in this community store. Also wholesale bakery and organic farm projects. Website was under construction at time of writing.
Contact: David Hoyle
Email: phoenix@findhorn.org
Address: The Park, Findhorn Bay, Moray, Highland IV36 3TZ, Scotland
Tel: 01309 690110
Fax: 01309 690933

Pillars of Hercules Farm Shop
www.pillars.co.uk

Specialities: Farm shop; box scheme
Website: Deliveries in Fife and Edinburgh. Grown on own farm or locally sourced, all fruit vegetables, eggs and meat are organically produced. There is a catalogue of products stocked with prices. Website gives an illustrated account of the farm and the farm shop.
Contact: Bruce Bennett
Email: mail@pillars.co.uk
Address: Pillars of Hercules, Falkland, Fife KY15 7AD, Scotland
Tel: 01337 857749

Pimhill Farm
www.pimhillorganic.co.uk

Specialities: Farm shop; café; box scheme
Website: Mixed organic farm. Vegetable box scheme operates locally. Pimhill Farm has been farming organically for 50 years. There is an award winning farm shop, café, and organic kitchen with in-house chef. The website gives details of the farm, the dairy, horticulture, the shop, café, box scheme, education, the herb garden and the threat of GMOs. There is also a newsletter.
Links: Yes
Contact: Ginny Mayall
Email: info@pimhillorganic.co.uk
Address: Harmer Hill, Shrewsbury, Shropshire SY4 3DY
Tel: 01939 290342
Fax: 01939 291156

Pitrok
www.pitrok.co.uk

Specialities: Bodycare; deodorants
Website: A range of natural body deodorants, both fragranced and unfragranced. All are based on mineral salts with very effective bacteriostatic action. Other products include a tongue cleaner and a remedy for snoring noise. On-line ordering facility and information for overseas customers. The range is also stocked by most independent healthfood shops. Website explains how the natural deodorant works.
On-line ordering: Yes
Export: Yes
Contact: Nigel Worthington
Email: info@pitrok.co.uk
Address: PO Box 1416, London W6 9WH
Tel: 020 8563 1120
Fax: 020 8563 9987

Pizza Organic
www.pizzapizza.co.uk

Speciality: Pizza
Website: Various branches in the south east and central England, all accredited

to the Soil Association. Website was under construction at time of writing
Tel: 020 8397 5556

Plamil Organics
www.plamilfoods.co.uk

Specialities: Dairy-free; gluten-free; chocolate; vegan
Website: Dairy and gluten-free organic chocolates, egg-free mayonnaise and sandwich spread. Website provides information pages on milk alternatives, mayonnaise, organic and soya chocolate and carob. There are also recipes. A booklet, *Vegan Case Histories*, is available for mail order sale.
Links: Yes
Contact: Arthur Ling
Email: plamil@veganvillage.co.uk
Address: Plamil House, Bowles Well Gardens, Folkestone, Kent CT19 6PQ
Tel: 01303 850588
Fax: 01303 850015

Planet Organic
www.planetorganic.com

Speciality: Supermarket
Website: The original organic and natural food supermarket. Over 8,000 products are carried. There is a mail order service. The website was under construction at time of writing
Email: sales@planetorganic.com
Address: 42 Westbourne Grove, London W2 5SH
Tel: 020 7221 7171
Fax: 020 7727 8547

Planet Vision.co.uk
www.planetvision.co.uk

Specialities: Clothing; baby wear
Website: Mail order service for men's, women's and baby clothing. Casual clothes are made of organic cotton, hemp and linen. Also supplies umbrellas and watches made entirely of recycled products. Product ranges shown on the website.
Address: 44 Parkway, Camden, London NW1 7AH
Tel: 020 7713 5575
Fax: 020 7713 5696

Plants for a Future
www.pfaf.org

Specialities: Vegan horticulture; gardening; permaculture; resource
Website: Research and information centre, promoting ecologically sustainable vegan organic permaculture. There is a mail order service for plants. On the website there is an introductory leaflet. There are details of the major project of a database of over 7,000 useful plant species with information on edible, medicinal and other uses together with cultivation and habitat requirements.
Links: Yes
Catalogue: Yes
Email: webmaster@pfaf.org
Address: Blagdon Cross, Ashwater, Beaworthy, Devon EX21 4DF
Tel: 01409 211694

Plaskett Nutritional Medicine College
www.pnmcollege.com

Specialities: Nutrition; courses; practitioner training
Website: Specialists in distance learning but also offer optional in-class and clinical training, A fully accredited practice diploma may be acquired by distance study supported by a tutor and occasional attendance at London weekends and annual clinical workshops. Website gives diary dates, on-line or printable application form and details of fees.
Contact: Dr Lawrence Plaskett
Email: Lgplaskett@aol.com
Address: 14 Southgate Chambers, Launceston, Cornwall PL15 9DY
Tel: 01566 773731
Fax: 01566 8630

Positive Health Magazine
www.positivehealth.com

Specialities: Magazine; complementary medicine, resource; directory
Website: This magazine on complementary medicine covers acupuncture, aromatherapy, asthma, bodywork, herbal and Chinese medicine, massage, nutrition, reflexology and yoga. The large website provides over 1,000 searchable articles, a research database, book reviews, forthcoming events, and an on-line directory of courses, therapists, suppliers and services. There is also a message board.
Links: Yes
Contact: Mike Howell
Email: admin@positivehealth.com
Address: Positive Health Publications, 51 Queen Square, Bristol BS1 4LH
Tel: 0117 983 8851
Fax: 0117 908 0097

Potions & Possibilities
www.potions.co.uk

Specialities: Aromatherapy products; bodycare; cosmetics
Website: A small Suffolk based company which offers bodycare products for the treatment of common ailments from its website catalogue.
On-line ordering: Yes
Catalogue: Mail order catalogue
Contact: Julie Foster
Email: enquiries@potions.co.uk
Address: The Forge, Bredfield, Woodbridge, Suffolk IP13 6AE
Tel: 01394 386161
Fax: 01394 384392

Potter's Herbal Medicines
www.pottersherbals.co.uk

Specialities: Herbal remedies; special diet products
Website: Manufacturers and distributors of fine herbal medicines and dietary products for almost 200 years. The website provides information on herbal medicine products categorised under the headings of mature, active, male and female. There are directories of ailments and herbal medicines which are cross referenced for easy access. There is also an FAQ page. Herbal medicines may be ordered on-line.
On-line ordering: Yes
Contact: Christine Morgan
Email: info@pottersherbals.co.uk
Address: Leyland Mill Lane, Wigan, Lancashire WN1 2SB
Tel: 01942 405100
Fax: 01942 820255

Power Health Products Ltd
www.power-health.com

Specialities: Food supplements; vitamins; bodycare; cosmetics; aromatherapy products
Website: A most comprehensive range of vitamin supplements, natural cosmetics, hair, skin, body sculpture and aromatherapy products. Own label and contract manufacturing facilities. Website gives company profile, full product listing and prices with on-line ordering facility. There are also information pages.
On-line ordering: Yes
Links: Yes
Contact: Vicky McIver
Address: 10 Central Avenue, Airfield Estate, Pocklington, Yorkshire YO42 1NR
Tel: 01759 302595
Fax: 01759 304286

Procter Rihl
www.procter-rihl.com

Speciality: Architects
Website: Multi-disciplinary studio which also specialises in green architectural design. It has designed several solar houses and also specialises in the use of natural light.
Email: info@procter-rihl.com
Address: 63 Cross Street, London N1 2BB

Tel: 020 7704 6003
Fax: 020 7688 0478

Provender Delicatessen
www.provender.net

Specialities: Delicatessen; groceries; dairy; juices; frozen vegetables; ice cream
Website: Delicatessen specialises in farmhouse cheeses and other fine food and drink from the West Country. There is on-line shopping. A number of recipes are also given.
On-line ordering: Yes
Credit cards: Visa; Mastercard
Export: Yes
Contact: Roger Biddle
Email: sales@provender.net
Address: 3 Market Square, South Petherton, Somerset TA13 5BT
Tel: 01460 240681
Fax: 01460 240681

Providence Farm Organic Meats
www.providencefarm.co.uk

Specialities: Meat; geese; poultry
Website: Organic pork, beef and lamb as well as organic poultry, geese and guinea fowl. On the website there is a price list for the various selection boxes of meat available. Orders may be placed by email, telephone or fax.
Credit cards: Visa; Mastercard
Contact: Ritchie & Pammy Riggs
Email: info@providencefarm.co.uk
Address: Crosspark Cross, Holesworthy, Devon EX22 6JW
Tel: 01409 254421
Fax: 01409 254421

The Pure H2O Company
www.pureh20.co.uk

Speciality: Water purifiers
Website: Website provides information on the purifiers and their health benefits together with further reading suggestions. There is also research material, a water comparison chart and details of the purifying process. A list of retail outlets is provided.
Email: info@pureh20.co.uk
Address: Unit 5, Egham Business Village, Crabtree Road, Egham, Surrey TW20 8RB
Tel: 01784 221188
Fax: 01784 221182

Pure Focus Lutein Spray
www.eyesight.nu

Specialities: Herbal remedies; eyecare
Website: Products for treatment of macular degeneration and other eye problems including glaucoma, cataracts, conjunctivitis, VDU eyestrain and squint. Personal action plans may be obtained by emailing details of eye condition.
Contact: Robert Redfern
Email: info@eyesight.nu
Address: Wholesale Health, 9 Stones Manor Lane, Hartford, Cheshire CW8 1NU
Tel: 0870 241 4237
Fax: 0870 122 8538

Pure Organic Food Ltd
www.pureorganicfoods.co.uk

Specialities: Meat; poultry; box scheme; wholesaler

Website: Leading wholesaler and distributor of organic meat and poultry throughout the UK. Supplies retail and catering outlets. On-line ordering is available for individual customers and there are introductory box selections.
On-line ordering: Yes
Credit cards: Visa; Mastercard
Contact: John Streeter
Email: enquiries@pureorganicfoods.co.uk
Address: Unit 5C, East Lands Industrial Estate, Leiston, Suffolk IP16 4LL
Tel: 01728 830575
Fax: 01728 833660

Pure Organics Ltd
www.pureorganics.co.uk

Specialities: Frozen foods; ready meals
Website: Products are available in supermarkets, healthfood shops, independent and organic retailers. They specialise in convenient organic foods which appeal to children. Product and stockist details are given on the website together with a summary of the company's history and aims. There is also a link to The Kids Organic Club (see separate entry).
Links: Yes
Contact: Pauline Stiles
Email: mail@pureorganics.co.uk
Address: Stockport Farm,
Stockport Road, Amesbury,
Wiltshire SP4 7LN
Tel: 01980 626263
Fax: 01980 626264

The Pure Wine Company
www.purewine.co.uk

Specialities: Wine; vegetarian & vegan
Website: Organic, vegetarian and vegan wines, all clearly marked, from around the world. Mixed case selections available. Complete UK coverage and also supplies to the retail trade.
Links: Yes
On-line ordering: Yes
Credit cards: Most major cards
Contact: Jim White
Email: service@purewine.co.uk
Address: Unit 18,
Browning Industrial Estate,
Respryn Road, Bodmin,
Cornwall PL31 1DQ
Tel: 01208 79300
Fax: 01208 79393

Queenswood Natural Foods
www.queenswoodfoods.co.uk

Speciality: Wholesaler
Website: Delivered wholesaler to the natural food trade operating throughout southern England. Website lists organic bulk commodities, branded goods, supplements, remedies and toiletries, frozen and chilled foods.
Contact: Terry Davies
Email: sales@queenswoodfoods.co.uk
Address: 2 Robins Drive,
Apple Business Park,
Bridgwater, Somerset TA6 4DL
Tel: 01278 423440
Fax: 01278 445826

Rachel's Organic Dairy
www.rachelsdairy.co.uk

Specialities: Dairy; yogurts
Website: Dairy products made from fresh liquid organic milk, organic fruits, and organic sugar. No thickeners, artificial flavours, colour, preservatives or stabilisers. Dairy products are listed and illustrated on the website with descriptions. A list of wholesale and retail outlets is provided. There is also a recipe page and an FAQ page.
Contact: Lindsay Collin &
Margaret Oakley
Email: enqs@rachelsdairy.co.uk
Address: Unit 63,
Glanyrafon Industrial Estate,
Aberystwyth,
Ceredigion SY23 2AE, Wales
Tel: 01970 625805
Fax: 01970 626591

The Ramblers Association
www.ramblers.org.uk

Specialities: Walking; organisation
Website: A very large website with general information, news, campaigns, walking information, features, groups, publications and events. There is a search facility. There is also an on-line membership application form.
Email: ramblers@london.ramblers.org
Address: 2nd Floor, Camelford House, 87-90 Albert Embankment, London SE1 7TW
Tel: 020 7339 8500
Fax: 020 7339 8501

Rapha UK Ltd
www.rapha.com

Speciality: Food supplements
Website: Manufacturer and distributor of a range of 'food state' supplements, unique in that they supply a strong foundation of nutrition in the same form as organic food. Website was under construction at time of writing.
Contact: Paul Geoghegan
Email: rapha@rapha.com
Address: I Stables Oddfellow, Walwyn Road, Colwall, Worcestershire WR13 6QW
Tel: 01684 541262
Fax: 01684 541577

Rare Breeds Survival Trust
www.rare-breeds.com

Specialities: Organisation; conservation; farming; livestock
Website: National charity dedicated to the preservation and conservation of rare breeds of farm livestock. Website sets out mission statement of the organisation. The website gives details of shows, sales, accredited butchers, approved centres, livestock sales and support groups. There is also a gene bank appeal to help save rare breeds of farm animals. A membership application form is available on the site.
Contact: Richard Lutwyche
Email: postmaster@rare-breeds.com
Address: National Agriculture Centre, Kenilworth, Warwickshire CV8 2LG
Tel: 024 7669 6551
Fax: 024 7669 6706

Rasanco Ltd
www.rasanco.com

Speciality: Food industry supplier
Website: Organic ingredients for manufacturers of food and drink: fruit, vegetables, starches, pulses, seeds, syrups, nuts, oils, spices, herbs, fats, cocoa, pasta, dairy and egg. Maintains a database of worldwide organic suppliers for specialist organic ingredients. Website gives details of the organic products available.
Contact: Russell Smart
Email: info@rasanco.com
Address: Old Estate Office, Sutton Scotney, Hampshire SO21 3JN
Tel: 01962 761935
Fax: 01962 761860

Ray Munn
www.raymunn.com

Specialities: Paints; paint remover
Website: Imports Becker range of latex based paints and also sells Homestrip water-based paint stripper. There is an on-line order form on the website.
Email:
enquiries@raymunn.freeserve.co.uk
Address: 861 Fulham Road, London SW6 5HP
Tel: 020 7736 9876
Fax: 020 7384 3273

Reading Scientific Services Ltd
www.rssl.com

Specialities: Consultancy; research
Website: A leading multi-disciplinary science based consultancy providing research, analysis and technical consultancy services to the food and natural product industries. Services include verification of authenticity of natural products, toxicity screen, pesticide testing, taints and off-flavours analysis, claims substantiation and nutritional analysis. There is a company profile on the website with details of services and quality. E-bulleting service available entitled Food-e-News.
Links: Yes
Contact: Alison Jeffs
Email: enquiries@rssl.com
Address:
Lord Zuckerman Research Centre,
Whiteknights, PO Box 234,
Reading, Berkshire RG6 6LA
Tel: 0118 986 8541
Fax: 0118 986 8932

Real Nappy Association
www.realnappy.com

Specialities: Organisation; nappies; babycare
Website: Membership organisation of companies selling washable nappies. Gives information on re-useable cloth nappies. The website provides information on nappy facts, washing instructions, where to buy, local authorities, chat rooms and press releases. There is a useful contact page with regional addresses and a membership application form.
Links: Yes
Address: PO Box 3704,
London SE26 4RX
Tel: 020 8299 4519

Really Healthy Company Ltd
www.healthy.co.uk

Speciality: Food supplements
Website: European supplier of innovative and effective health products from the USA. Organic food supplements based on algae harvested in Oregon. Website provides full information on the products with an on-line ordering facility, There are pages on tips and books.
Links: Yes
On-line ordering: Yes
Credit cards: Most major credit cards
Email: sales@healthy.co.uk
Address: 4 Tanner Street,
London SE1 3LD
Tel: 020 8480 1000
Fax: 020 8480 1010

Reaseheath College
www.reaseheath.ac.uk

Specialities: Courses; gardening; farming; agriculture
Website: Provides courses in organic gardening, agriculture, the countryside, food and dairy technology, and horticulture amongst others. Full prospectus available from the website
Links: Yes
Address: Nantwich,
Cheshire CW5 6DF
Tel: 01270 625131
Fax: 01270 625660

Recoup
www.recoup.org

Specialities: Organisation; recycling
Website: Charity which promotes and facilitates the recycling of plastic containers and provides educational material. On the website there is information about Recoup, factsheets, education and promotion, end products and membership details.
Email: enquiry@recoup.org
Address: 9 Metro Centre,
Welbeck Way, Woodston,
Peterborough,
Cambridgeshire PE2 7WH
Tel: 01733 390021
Fax: 01733 390031

Recycled Paper Supplies
www.recycled-paper.co.uk

Specialities: Recycled paper; printing; stationery
Website: Specialist suppliers of quality recycled paper and stationery products by mail order. Product range covered on the website includes writing paper, envelopes and card, copier and ink-jet papers, art and craft materials, wrapping papers, self-adhesive and gummed labels. The company also prints personal and wedding stationery. There are information pages on recycling. There is also an on-line catalogue with on-line ordering facility.
Links: Yes
On-line ordering: Yes
Credit cards: Visa; Mastercard; Switch; Solo; JCB
Catalogue: Yes (gratis)
Contact: Steve & Susan Hammet
Email: orders@recycled-paper.co.uk
Address: Gate Farm, Fen End, Kenilworth, Warwickshire CV8 1NW
Tel: 01676 533832
Fax: 01676 533832

Regency Mowbray Company Ltd
www.regencymowbray.co.uk

Specialities: Food industry supplier; vegetarian
Website: Manufacturers of flavourings, colourings, fruit & chocolate products, emulsifiers and stabilisers, some of which are organic and vegetarian. A list of agents and distributors is given.
Email: sales@regencymowbray.co.uk
Address: Hixon Industrial Estate, Hixon, Staffordshire ST18 0PY
Tel: 01889 270554
Fax: 01889 270927

Regent Academy
www.regentacademy.com

Specialities: Aromatherapy; reflexology; counseling; parenting; feng shui; courses
Website: Introductory home-study courses in a variety of health and lifestyle subjects. Website gives details of the individual courses and their modules and course materials. Fees are shown and enrolment may be made with a credit card over the phone or through the printable enrolment form.
Email: info@regentacademy.com
Address: 155 Regent Street, London W8 4PX
Tel: 0800 378 281
Fax: 020 7287 3348

Reiki Association
www.reikiassociation.org.uk

Specialities: Reiki; organisation
Website: Website provides a statement of identity for the association, discusses what is reiki and defines the usui system of reiki, gives reiki precepts and a history of the association together with the code of ethics.
Email: katejones@reikiassociation.org
Address: Cornbrook Bridge House, Clee Hill, Ludlow, Shropshire SY8 3QQ
Tel: 01584 891 197

Resurgence Magazine
www.resurgence.org

Specialities: Magazine; ecology
Website: Magazine covering ecology, L.E.T.S., the organic movement, sustainability, art and culture, book reviews etc. The website provides a selection of recent articles on-line. There is also a green events listing, information on advertising and media resources. Books and back issues of the magazine may be purchased. Subscription details are available.
Links: Yes
Contact: Jeannette Gill
Email: subs.resurge@virgin.net
Address: Rocksea Farmhouse, St Mabyn, Bodmin, Cornwall PL30 3BR
Tel: 01208 841824

The Retreat Association
www.retreats.org.uk

Specialities: Retreats; organisation
Website: The association is a charity comprising 6 British Christian retreat groups. The website lists the groups and provides links to websites of individual retreat houses throughout the country listed by region. It also provides a mission statement and information on publications and the *Retreats* journal.
Email: info@retreats.org.uk
Address: Central Hall,
256 Bermondsey Street,
London SE1 3UJ
Tel: 020 7357 7736

The Retreat Company
www.retreat-co.co.uk

Specialities: Retreats; holidays
Website: The company promotes, organises and facilitates time-out activities, workshops, mini-breaks and holidays for personal transformation and the maintenance and balance of well-being. It covers yoga, breathing, meditation, healing, relationships, family, expression, seers and saints, ayurveda, personal development and healing therapies. Website has details of what's on and where to go and information on the *Retreat Directory* which is published 3 times a year.
Email: timeout@retreat-co.co.uk
Address: The Manor House,
Kings Norton, Leicestershire LE7 9BA
Tel: 0116 259 9211
Fax: 0116 259 6633

Revital Stores
www.revital.com

Specialities: Toiletries; babycare; food supplements; vitamins & minerals; herbal remedies; homoeopathy
Website: On-line health and wellness store. Over 10,000 products are available for on-line purchasing through the electronic catalogue.
On-line ordering: Yes
Credit cards: Most major credit cards
Email: enquire@revital.com
Address: 3a, The Colonnades,
123 - 151 Buckingham Palace Road,
London SW1 9RZ
Tel: 0800 252 875
Fax: 020 7976 5529

Richard Guy's Real Meat Company
www.realmeat.co.uk

Specialities: Meat; poultry
Website: Site explains the difference between terms such as 'organic' and 'free range' and how supermarkets might exploit them. On-line ordering facility including introductory hamper. Organic beef, pork, lamb, poultry, ham, bacon and sausages are available.
On-line ordering: Yes
Credit cards: Most major cards
Email: richard@realmeat.co.uk
Address: Warminster,
Wiltshire BA12 0HR
Tel: 01985 840562
Fax: 01985 841005

Riverford Organic Vegetables
www.riverford.co.uk

Specialities: Box scheme; vegetables; dairy; farm shop
Website: Largest organic vegetable producers in the UK. Wholesale, local sales and home deliveries. Award winning organic box scheme delivered throughout the South West. Details of the farm are given on the website. There is a postcode search to check whether there is box scheme delivery available in a particular area. The site also gives information on the dairy as well as farm news, recipes and conservation.
Email: mail@riverford.co.uk
Address: Wash Barn, Buckfastleigh,
Devon TQ11 0LD
Tel: 01803 762720
Fax: 01803 762718

Rockland Corporation - TRC
www.trceurope.com

Specialities: Food supplements; household products
Website: Plant derived colloidal mineral products, liquid formulations, encapsulation, mining and earth friendly cleaning chemicals. US based website gives UK prices and product descriptions with full ingredients and supplement facts.
Contact: Don Stables
Email: trceurope@aol.com
Address: Unit A, Grovebell Industrial Estate, Wrecclesham Road, Farnham, Surrey GU10 4PL
Tel: 01252 719402
Fax: 01252 719415

Rococo Chocolates
www.rococochocolates.com

Specialities: Chocolate; vegetarian & vegan
Website: Handmade organic chocolates and a range of really high quality chocolates for vegans, vegetarians and serious chocolate lovers. Website gives background information on real chocolate. There is an on-line catalogue of chocolates for ordering.
Links: Yes
On-line ordering: Yes
Credit cards: All major credit cards
Catalogue: Yes
Contact: Alex
Email: sales@rococochocolates.com
Address: 321 King's Road, London SW3 5EP
Tel: 020 7352 7360
Fax: 020 7352 7360

Rocombe Farm Fresh Ice Cream Ltd
www.rocombefarm.co.uk

Specialities: Ice cream; frozen yogurt; fruit sorbets
Website: Organic cream made from organic dairy milk in Devon. The site provides a history of how the company expanded from a small shop to nationwide availability in fine food shops and restaurants and some branches of the large supermarket chains. A list of current stockists is being compiled.
Contact: Peter Redstone
Email: info@rocombefarm.co.uk
Address: Old Newton Road, Heathfield, Newton Abbot, Devon TQ12 6RA
Tel: 01626 834545
Fax: 01626 835777

Rose Blanc Rouge
www.rose-blanc-rouge.com

Specialities: Wine; beers; juices
Website: A selection of fine organic wines, beers and juices. The website provides an explanation of organic wine and an on-line catalogue with details of production including grape type, soil and fining. There are tasting notes.
Contact: Muriel Chatel & Stephan Gervois
Email: contact@rose-blanc-rouge.com
Address: 2 Bedale Street, London SE1 9AL
Tel: 020 7403 0358
Fax: 020 7403 0358

Rowse Honey Ltd
www.rowsehoney.co.uk

Specialities: Honey; maple syrup; importer
Website: Imported organic honey from Australia, New Zealand, Argentina, Mexico and Turkey. Also imported organic maple syrup from Canada. Rowse is the major supplier of organic honey to supermarkets in the UK. The organic honey range has a special section on the website. There are recipe pages and a useful FAQ on honey.
Contact: Stuart Bailey
Email: info@rowsehoney.co.uk
Address: Moreton Avenue, Wallingford, Oxfordshire OX10 9DE
Tel: 01491 827400
Fax: 01491 827434

Rudolf Steiner School Kings Langley
www.rudolfsteiner.herts.sch.uk

Speciality: School
Website: One of nearly 500 schools throughout the world educating children according to the curriculum devised by Rudolf Steiner. Website provides background information on the school and Steiner principles and philosophy, and also gives details of education, the kindergarten, lower school and upper school. There are links to other Steiner education sites.
Email: info@rudolfsteiner.herts.sch.uk
Address: Langley Hill, Kings Langley, Hertfordshire WD4 9HG
Tel: 01923 262 505
Fax: 01923 270 958

Rush Potatoes
www.rushpotatoes.co.uk

Speciality: Potatoes; wholesaler
Website: Specialises in growing out of season organic potatoes for import to the UK. Website provides company profile and interactive pages for bulk potato offers and bids. There is a good practice declaration for potential suppliers.
Contact: Murray Hogge
Email: sales@rushpotatoes.co.uk
Address: 1 Harbour Exchange Square, London E14 9GE
Tel: 020 7363 7887
Fax: 020 7363 7870

S.A.D. Lightbox Co Ltd
www.sad.uk.com

Specialities:
Seasonal affective disorder; light therapy
Website: Manufacturers of light therapy equipment to deal with SAD, sleep disorders, jetlag and ME. Website gives illustrated details of the range of lightboxes, both wooden and plastic, which can all be purchased on-line.
On-line ordering: Yes
Catalogue: Free information pack
Email: info@sad.uk.com
Address: 19 Lincoln Road, Cressex Business Park, High Wycombe, Buckinghamshire HP12 3FX
Tel: 01494 448 727
Fax: 01494 527 005

Sainsbury's
www.sainsbury.co.uk

Specialities: Supermarket; home delivery; groceries; wines; baby foods
Website: Part of general on-line catalogue available in selected areas. Organic can be specified in screening box of main catalogue. Browse and order facilities. Help line for internet and access problems. Organic wines can be ordered on-line for collection at the Calais store.
On-line ordering: Yes
Credit cards: Major cards
Address: Sainsbury's To You, St James's House, 1-5 Wilder Street, Bristol BS2 8QY
Tel: 0845 3012020

St Peter's Brewery
www.stpetersbrewery.co.uk

Speciality: Beer
Website: Independent brewery of organic and other ales in historic location. Website provides descriptions of beers, bottled and cask, as well as distributors and outlets. On-line ordering facility. Links to other relevant sites.
Links: Yes
On-line ordering: Yes
Email: beers@stpetersbrewery.co.uk
Address: St Peter's Hall, St Peter's, South Elmham, Bungay, Suffolk NR35 1NQ
Tel: 01986 782322
Fax: 01986 782505

Salvo
www.salvo.co.uk

Specialities: Building materials; architectural salvage
Website: Encourages the use of reclaimed building materials through a newsletter and directory. Provides topical information on auctions and demolitions. Information pack available giving information on where to buy antique and reclaimed items on a regional basis. There is a similar pack for France and Belgium. The website gives details of items for sale, auctions, publications, wants listings and services.
Email: tk@salvoweb.com
Address: PO Box 333, Cornhill on Tweed, Northumberland TD12 4VJ
Tel: 01890 820333
Fax: 01890 820499

Sam-I-Am
www.nappies.net

Specialities: Nappies; baby products
Website: Environmentally friendly cotton nappies and accessories. An order form is available on the website which may be printed out and posted. There is an information section which covers health, environment, economics, convenience and the dangers of disposable nappies.
Credit cards: Most major cards
Address: 4 Sharon Road, Chiswick, London W4 4PD
Tel: 020 8995 9204

Sarah Wrigglesworth Architects
www.swarch.co.uk

Speciality: Architects
Website: Specialist in buildings which use ordinary, inexpensive or recycled materials in innovative and exciting ways. Emphasis on not contributing to global warming and choosing materials which are low in energy design and the use of resources. The website has section headings on projects, publications and what's new.
Email: email@swarch.co.uk
Address: 9/10 Stock Orchard Street, London N7 9RW
Tel: 020 7607 9200
Fax: 020 7607 5800

Sauce Organic Diner
www.sauce-organicdining.co.uk

Specialities: Restaurant; juice bar
Website: Organic diner with kid's menu and take away service. Licensed to sell organic beers and wines and juice bar. Website was under construction at the time of writing.
Contact: Karen & Ross Doherty
Email: dining@sauce.prestel.co.uk
Address: 214 Camden High Street, London NW1 8QR
Tel: 020 7482 0777

Savant Distribution
www.savant-health.com

Specialities: Distributors; food supplements; bodycare; health equipment
Website: UK importers and distributors of a number of branded lines. The site brings together the entire catalogue together with a wide range of information on the products themselves and the processes involved in their manufacture. Catalogue covers nutritional oils, dietary supplements, biocare supplements, herbal products, products for pets, health equipment and books, all of which may be ordered on-line.
Links: Yes
On-line ordering: yes
Export: Yes
Contact: Ian Richardson
Email: info@savant-health.com
Address: 15 Iveson Approach, Ireland Wood, Leeds, West Yorkshire LS16 8LJ
Tel: 0113 230 1993
Fax: 0113 230 1915

J. Savory Eggs
www.broadland.com/highfield

Specialities: Eggs; bed & breakfast; holidays
Website: Egg producer. There is also farmhouse bed and breakfast, details of which may be found on the website.
Contact: Elizabeth Savory
Email: jegshighfield@onet.co.uk
Address: Highfield Farm, Great Ryburgh, Fakenham, Norfolk NR21 7AL
Tel: 01328 829249
Fax: 01382 829422

School of Homoeopathy
www.homeopathyschool.com

Specialities: Homoeopathy; courses
Website: Training for those wishing to become practising homoeopaths with courses on campuses in Devon and New York as well as correspondence tuition. Website provides an introduction to the school, its faculty and the various aspects of study. It also introduces a new service that provides open information on some of the many homoeopathic provings carried out by the school and its faculty.
Catalogue: Prospectus
Contact: Stuart Gracie
Email: stuart@homeopathyschool.com
Address: Orchard House, Merthyr Road, Llanfoist, Abergavenny, Monmouthshire NP7 9LN, Wales
Tel: 01873 856 872
Fax: 01873 858 962

School of Meditation
www.schoolofmeditation.org

Specialities: Meditation; courses
Website: Questions on meditation are answered on the website which also gives information on the nature of the school and what and how it teaches. There is a school calendar which gives details and times of public meetings at which short talks are given and questions may be put.
Address: 158 Holland Park Avenue, London, W11 4UH
Tel: 020 7603 6116
Fax: 020 7603 6116

Schumacher UK
www.schumacher.org.uk

Specialities: Sustainable development; conservation; organisation
Website: Promoting human-scale sustainable development in Britain and abroad, the society plays a major role in encouraging the spread of ecological and spiritual values. The vision is informed by a sense of the whole, the connectedness of the parts, a commitment to sustainability and a reverence for the sacred. Website provides information on membership, the Schumacher lectures, awards and briefings, the book service and E.F. Schumacher himself.
Email: schumacher@gn.apc.org
Address: Create Environment Centre, Smeaton Road, Bristol BS1 6XN
Tel: 0117 903 1081
Fax: 0117 903 1081

Scottish Agricultural College
www.sac.ac.uk/organic-farming

Specialities: Research; agriculture; advice; courses; farming; consultancy
Website: Provision of advice as part of the Scottish Organic Aid Scheme, the website provides information on the farming centre, organic farms, research and development, education and training, and cereal growing. An organic conversion information pack is available.
Contact: David Younie
Email: d.younie@ab.sac.ac.uk
Address: Craibstone Estate, Bucksburn, Aberdeen, Grampian AB21 9YA, Scotland
Tel: 01224 711072
Fax: 01224 711293

Scottish School of Herbal Medicine
www.herbalmedicine.org.uk

Specialities: Courses; massage; aromatherapy; herbalism; organisation
Website: A non-profit making organisation dedicated to the furthering of herbal knowledge and education. Website gives details of teaching, library, events, and research. There is a listing of herbalists in Scotland by area. Full details of courses, including by correspondence, are on the website.
Links: Yes
Email: sshm@herbalmedicine.org.uk
Address: 6 Harmony Row, Glasgow, Strathclyde G51 3BA, Scotland
Tel: 0141 401 8891
Fax: 0141 401 8889

Scullions Organic Supplies
www.scullionsorganicsupplies.co.uk

Specialities: Fruit; vegetables; meat; dairy; groceries; toiletries
Website: Retail business offering all organic products. Website gives a full on-line listing of products available together with prices. Orders can be emailed. There is free delivery for orders over £25.00 within a 20 mile radius.
Contact: Joyce Scullion
Email: orders@scullionsorganicsupplies.co.uk
Address: 143 Stamperland Gardens, Clarkston, Glasgow, Strathclyde G76 3LJ, Scotland
Tel: 0141 638 6200
Fax: 0141 881 1373

Sedlescombe Vineyard
www.tor.co.uk/sedlescombe

Specialities: Wines; vineyard; beers; cider; juices
Website: Organic vineyard planted in 1979 and producer of organic English wines. Also importers of organic beer from Bavaria and organic wine from France. There is a secure on-line organic wine shop. Tasting notes provided. There is also information on organic wines, the history of Sedlescombe, and working on organic farms.
On-line ordering: Yes
Contact: Roy Cook
Email: sales@organicwines.netscapeonline.co.uk
Address: Cripp's Corner, Sedlescombe, Nr Robertsbridge, East Sussex
Tel: 01580 830715
Fax: 01580 830122

Seeds of Change
www.seedsofchange.co.uk

Specialities: Pasta; sauces; soups; ketchup
Website: Seeks to protect the planet's biodiversity and promote organic agricultural practices by offering a diverse range of open pollinated 100% organic products. Products are available nationally. The site includes historical and organic information as well as recipes.
Links: Yes
Contact: Lynn Simpson
Address: Freeby Lane, Waltham on the Wolds, Leicestershire LE14 4RS
Tel: 0800 952 0000

Seeds of Italy
www.seedsofitaly.com

Specialities: Seeds; vegetable seeds; herbs gardening
Website: Wide range of seeds including organic range certified in Bologna. On-line ordering facility and catalogue request facility.
Links: Yes
On-line ordering: Yes
Catalogue: Yes (gratis)
Email: grow@italianingredients.com
Address: 260 West Hendon, Broadway, London NW9 6BE
Tel: 020 8930 2516
Fax: 020 8930 2516

Selfheal School
www.selfhealschool.co.uk

Specialities: Herbal therapies; natural healing; courses
Website: The school offers training at several levels: a short home-study course, weekend workshops and a 12 month course based on home-study and 3 attendance weekends. The courses do not lead to a professional qualification but are intended to help people, their friends and family.
Catalogue: Free prospectus
Email: selfheal@aol.com
Address: The Cabins, Station Warehouse, Station Road, Pulham Market, Norfolk IP21 4XF
Tel: 01379 608082
Fax: 01379 608201

September Organic Dairy
www.september-organic.co.uk

Speciality: Ice cream
Website: Makers of organic dairy ice creams which are available in independent stores throughout the country. Website gives information on the ice cream and its flavours, the farm and the people, suppliers of the ice cream by region, and other organic products available.
Email: enquiries@september-organic.co.uk
Address: Newhouse Farm, Almeley, Nr Kington, Herefordshire HR3 6LJ
Tel: 01544 327561
Fax: 01544 327561

Shared Interest Society
www.shared-interest.com

Specialities: Building society; financial services
Website: A co-operative lending society with about 8,500 members. A proportion of the members' savings is invested in helping Third World producers pay for the cost of producing their goods until the goods can be sold to consumers. Website has pages on investment, the clearing house, who's who, handbook, resources, corporate and news.
Email: post@shared-interest.com
Address: 25 Collingwood Street, Newcastle upon Tyne, Tyne and Wear NE1 1JE
Tel: 0845 840 9100
Fax: 0191 233 9110

Sharpham Partnership
www.sharpham.com

Specialities: Vegetables; dairy; wine
Website: Family company producing organic vegetables and dairy products on their own farm. The website gives details of the wines and cheeses available. There is also information on the two self-guided trails available on the estate.
Links: Yes
Contact: M Sharman
Email: info@sharpham.com
Address: Sharpham Estate, Ashprinton, Totnes, Devon TQ9 7UT
Tel: 01803 732203
Fax: 01803 732122

Sheepdrove Organic Farm
www.sheepdrove.com

Specialities: Meat; poultry
Website: A traditional mixed farm that is self sustaining. Soil Association certified organic chicken, turkey, lamb, beef and pork through farmers' markets, mail order and direct delivery service. There is on-line email registration for the regular newsletter. Education and conservation pages. Home delivery in vacuum packed insulated chill boxes. Price list is available on-line. On-line order form for emailing.
Contact: Peter Molesworth
Email: manager@sheepdrove.com
Address: Warren Farm Office, Sheepdrove, Lambourn, Berkshire RG17 7UU
Tel: 01488 71659
Fax: 01488 72677

Shepherdboy Ltd
www.shepherdboy.co.uk

Specialities: Confectionery; breakfast foods;
Website: Independent manufacturers of natural health food, fruit and nut snack bars, muesli and hemp oil capsules. Also organic muesli and sunflower bars. Products are available in all good healthfood shops. Website gives details of products, what's new and health tips.
Contact: David Revitt
Email: enquiries@shepherdboy.co.uk
Address: Healthcross House, Cross Street, Syston, Leicestershire LE7 2JG
Tel: 0116 2602 992
Fax: 0116 2693 106

The Shiatsu College of London
www.shiatsucollege.co.uk

Specialities: Shiatsu; courses; practitioner training
Website: This website covers a network of shiatsu schools throughout the UK. Courses are co-ordinated through the college. Courses range from the introductory to a three year practitioner training course.
Links: Yes
Address: Unit 62, Pall Mall Deposit, 126 Barbly Road, London W10 6BL
Tel: 020 8987 0208
Fax: 020 8987 0208

The Shieling
www.house-of-eden.co.uk

Specialities: Gluten-free; café
Website: An extensive range of over 80 gluten, wheat, milk, soya and egg-free foods are available by mail order. A list can be supplied on demand. Website gives brief illustrated details of the shop, café and gift shop.
Catalogue: List
Contact: Michelle Herd
Email: info@theshieling.co.uk
Address: 18 Dee Street, Banchory, Aberdeenshire AB31 5ST, Scotland
Tel: 01330 823278
Fax: 01224 749288

Simply Organic
www.simplyorganic.co.uk

Specialities: Babyfood; pasta sauces; soups
Website: A young Scottish company making a range of fresh organic foods including soups, pasta sauces and baby food. There is a listing of products on the website. The foods are available through some major supermarkets as well as specialist independent retailers and healthfood shops. Trade customers can register on-line for more detailed information.
Contact: Belinda Mitchell
Email: information@simplyorganic.co.uk
Address: Unit 21 - 22, Dryden Vale, Bilston Glen, Midlothian EH20 9HN, Scotland
Tel: 0131 448 0440

SimplyOrganic Food Company
www.simplyorganic.net

Specialities: Supermarket; baby food; vegetables; meat; wines; home delivery
Website: Over 1,900 organic products ranging through fruit and vegetables, meat, fish, dairy, groceries, infant care, personal care, home care, wines, beers and spirits, pet food and organic lifestyle. Order by telephone, fax or internet. Customers anywhere on the UK mainland can pick delivery day of their choice. On-line option to search for specific foods or browse through the aisles. Catalogues may be ordered on-line.
Credit cards: Visa; Mastercard; Delta; Switch
Catalogue: Yes
Contact: Simon Grant
Email: info@simplyorganic.net
Address: A62 - A64 New Covent Garden Market, London SW8 5EE
Tel: 0845 1000 444
Fax: 020 7622 4447

Simply Soaps
www.simplysoaps.com

Specialities: Soaps; bodycare; vegan
Website: A skincare company making fine soaps with ingredients sourced from around the world which are organic where possible. Website gives details of the soaps and their method of manufacture. Vegan products are listed. There is an on-line order form for printing and faxing.
Export: Yes
Contact: Jim Jones
Email: info@simplysoaps.com
Address: Brillig Rackheath Park, Rackheath, Norwich, Norfolk NR13 6LP
Tel: 01603 720869
Fax: 01603 722159

Skyros
www.skyros.com

Specialities: Holidays; courses
Website: Holistic holidays on Skyros, Greece and Ko Samet, Thailand. There is a printable booking form as well as an on-line booking form. A very wide range of courses is available and details, dates and venues are given on-line.
Catalogue: Free brochure
Email: connect@skyros.com
Address: 92 Prince of Wales Road, London NW5 3NE
Tel: 020 7284 3065

Slipstream Organics
www.slipstream-organics.co.uk

Speciality: Box scheme
Website: Box scheme for Cheltenham, Gloucester, Stroud and surroundings. The website gives basic information.
Links: Yes
Contact: Nick McCordall
Email: nick@slipstream-organics.co.uk
Address: 34A Langdon Road, Cheltenham, Gloucestershire GL53 7NZ
Tel: 01242 227273
Fax: 01242 227798

Slow Food
www.slowfood.com

Speciality: Eating
Website: Founded in Italy in reaction to the growth of the fast-food culture with a philosophy to promote the pleasures of the table, aiming to protect and celebrate food as a reflection of local culture, skill and tradition. Although there are 65,000 members there are still very few indeed in the UK. The international website provides information on the organisation in Italy and worldwide, events, recipes and membership details. There is on-line membership registration.
On-line ordering: Yes
Email: internationalinfo@slowfood.com
Tel: 01844 339 362

Society of Teachers of Alexander Technique (STAT)
www.stat.org.uk

Specialities: Alexander technique; organisation
Website: The largest regulatory body for the Alexander technique. Website provides background information on the technique and its founder. It gives information on courses and events, the technique as used in business and commerce, how to train as a teacher, books, articles, tapes and how to find a teacher.
Email: enquiries@stat.org.uk
Address: 129 Camden Mews, London NW1 9AH
Tel: 020 7284 3338
Fax: 020 7482 5435

Soil Association
www.soilassociation.org

Specialities: Organisation; certification; environment; sustainable development; directory
Website: The driving force behind the organic movement which provides Soil Association certification. Advice on organic production and processing, conversion and certification, and the UK organic market. The Soil Association exists to research, develop and promote sustainable relationships between the soil, plants, animals and people in order to produce healthy food and other products while protecting and enhancing the environment. It is at the forefront of the campaign against GM foods. Very extensive web site with information, library, questions answered, campaigns and forestry. There is also an on-line organic directory.
Contact: Dom Lane
Email: info@soilassociation.org
Address: Bristol House,
40-56 Victoria Street, Bristol BS1 6BY
Tel: 0117 929 0661
Fax: 0117 925 2504

Solar Century
www.solarcentury.co.uk

Specialities: Solar energy; alternative energy
Website: A solar PV solutions business which designs, installs, and maintains tailor-made solar PV systems for businesses, homes, industry and the public sector. If every suitable roof in the UK had PV, all electricity needs could be generated from solar power. Website gives information on the company and the technology.
Email: enquiries@solarcentury.co.uk
Address: Unit 5, Sandycombe Centre,
1-9 Sandycombe Road, Richmond,
Surrey TW9 2EP
Tel: 0870 735 8100
Fax: 0870 735 8101

Solar Sense
www.solarsense.co.uk

Specialities: Solar energy; alternative energy
Website: Company aims to provide easy-to-install modular components for the self-build and housing association-constructor markets. The website provides information on the different components of the system and gives specifications. Prices on application.
Links: Yes
Email: info@solarsense.co.uk
Address: Tree Tops, Sandy Lane,
Pennard, Swansea SA3 2EN, Wales
Tel: 01792 371 690
Fax: 01792 371 690

The Solar Design Company
www.solardesign.co.uk

Specialities: Alternative energy; solar energy
Website: Designs, installs and maintains solar water heating systems. The website covers solar water heating, efficient heating, alternative energy, past projects and company profile.
Email: chris@solardesign.co.uk
Address: 57 Wood Lane, Greasby,
Wirral, Merseyside CH49 2PU
Tel: 0151 606 0207
Fax: 08700 526977

Solar Trade Association
www.solartradeassociation.org.uk

Specialities: Organisation; alternative energy
Website: UK member of the association of businesses which manufacture, supply, and install solar energy systems. It provides a useful source of information. Site under construction at time of writing.
Email:
enquiries@solartradeassociation.org.uk
Address: The Energy Centre,
Davy Avenue, Knowlhill Road,
Milton Keynes,
Buckinghamshire MK5 8NG
Tel: 01908 442290
Fax: 0870 0529194

Solarsaver
www.solarsaver.co.uk

Specialities: Solar energy; alternative energy; wind energy
Website: Over 20 years experience in renewable energy, operating at the leading edge for an energy efficient environment in the commercial, industrial and domestic markets. Website has headings on the solar-saver house, solar energy kit, renewable energy systems, wind turbines, domestic user, business user, sun pipes, sun catchers, wind catchers and watermatic.
Email: action@solarsaver.co.uk
Address: 22 Clay Hill Road, Sleaford, Lincolnshire NG34 7TF
Tel: 01529 304 027
Fax: 01529 410 491

Solartwin Ltd
www.solartwin.com

Specialities: Alternative energy; solar energy
Website: Affordable freeze-tolerant solar water heating system. No costly environmentally-unfriendly anti-freeze required. Website provides sectionalised diagram of how the system works. Questions and answers, suitability, ease of installation, photos, diagrams, technical answers and questions, survey requirements are all covered. Prices are given for installed and self-install systems. There is a printable order form.
Email: hi@solartwin.com
Address: 15 King Street, Chester, Cheshire CH1 2AH
Tel: 01244 403 404
Fax: 01244 403 654

Solgar Vitamins
www.solgar.com

Specialities: Food supplements; vitamins
Website: Manufacturers of the award-winning gold label range of over 425 nutritional supplements. US based website gives company and product information and there is a search facility by county for UK stockists.
On-line ordering: Yes
Contact: Marie Kendal
Email: kendalm@solgar.com
Address: Aldbury, Tring, Hertfordshire HP23 5PT
Tel: 01442 890355
Fax: 01442 890366

Solid Floor
www.solidfloor.co.uk

Speciality: Flooring
Website: Extensive range of wood flooring with varied finishes, textures and colours. Website gives details of woods with photographs to show colours and finishes. There are two showrooms in London and one in Manchester. There is an email request form on the site for further information.
Catalogue: Brochure
Address: 53 Pembridge Hill, Notting Hill Gate, London W11 3HG
Tel: 020 7221 9166
Fax: 020 7221 8193

Soma Organic Smoothies
www.somajuice.com

Speciality: Smoothies
Website: Organic smoothies - Jungle Juice made from fruit with added spirulina, wheat and barley grasses, blue green algae and soya lecithin and Cool Bananas made from a tropical mixture of fruit and vanilla pods.
Contact: Karim Sams
Email: us@somajuice.com
Address: 22 Islingword Road, Brighton, East Sussex BN2 2SE
Tel: 01273 236826
Fax: 01273 236826

Somerset Farm Direct
www.somersetfarmdirect.co.uk

Specialities: Meat; poultry; game
Website: Traditionally reared lamb, poultry and game direct from the farm. Meat comes from animals that are well cared for and reared in traditional ways under non-intensive methods. On the website there is a product listing of the different meat selection boxes with prices for on-line ordering. Recipe ideas are also given on the site.
On-line ordering: Yes
Credit cards: All major credit & debit cards
Email: deawood@btinternet.com
Address: Bittescombe Manor, Wiveliscombe, Taunton, Somerset TA4 2DA
Tel: 01398 371387
Fax: 01398 371413

Somerset Levels Organic Foods
www.somersetorganics.com

Specialities: Meat; poultry; geese
Website: On-line ordering of organic meat. Mail order service with nationwide delivery. CD rom provides virtual farm tour. Beef, lamb, pork, duck, chicken, game, bacon, ham, turkey, goose and sausages and also organic farmed trout are available. There is a downloadable brochure.
On-line ordering: Yes
Credit cards: Most major cards
Catalogue: Brochure
Contact: Richard Counsell
Email: info@somersetorganics.co.uk
Address: Manor Farm, Rodney Stoke, Cheddar, Somerset BS27 3UN
Tel: 01749 870919
Fax: 01749 870919

Soup Dragon
www.soup-dragon.co.uk

Specialities: Wooden toys; games; children's clothing; fancy dress
Website: Unusual wooden toys, puzzles and games and also kids' clothing and fancy dress. The emphasis is on the traditional and the unusual. There is information on knitwear and baby toys, clothes and sit-and-ride toys, fairy dresses and rocking horses, dolls houses and fancy dress. Mail order is available and on-line ordering was under development at the time of writing.
Email: info@soup-dragon.co.uk
Address: 27 Topsfield Parade, Tottenham Lane, Crouch End, London N8 8PT
Tel: 020 8348 0224
Fax: 020 8348 9555

Source Naturals - Planetary Formulas
www.sourcenaturals.com

Specialities: Herbal remedies; food supplements
Website: Natural immune-support products. US based site which gives information pages on strategies for wellness, alternative healthcare information and lifestyle plans. There is also extensive product information. An email newsletter is available.
Contact: Jan McWhir
Email: earth.force@cableinet.co.uk
Address: Earthforce, 1 Upper Belmont Road, Bristol BS7 9DG
Tel: 0117 904 9930
Fax: 0117 904 9931

South Devon Organic Producers
www.sdopltd.co.uk

Speciality: Vegetables
Website: Markets a wide range of organic vegetables direct from the growers. The website was under construction at the time of writing.
Contact: Joanna Field
Email: sdop@farmersweekly.net
Address: Sharpham Estate, Ashprington, Totnes, Devon TQ9 7UT
Tel: 01803 762100
Fax: 01803 762755

Spice Direct
www.spicedirect.co.uk

Specialities: Herbal remedies; aromatherapy products; books
Website: Suppliers of aromatherapy and herbal products to the trade and to the public. The company is a web-based mail order company supplying affordable high quality products. The website provides a complete ordering facility and on-line catalogue as well as information on aromatherapy and herbalism, a newsletter, a plant guide and an A - Z of ailments. There is also a therapist directory.
Links: Yes
On-line ordering: Yes
Credit cards: Most major credit cards
Email: sales@spicedirect.co.uk
Address: Jaemar House, Mill Lane, Manningtree, Essex CO11 1DU
Tel: 0870 90 77 123
Fax: 01206 390119

Spirit of Nature
www.spiritofnature.co.uk

Specialities: Clothing; baby clothing; nappies; books; household products
Website: Natural clothing and environmentally friendly products. Over 1,000 items are available. On-line ordering is provided with helpful sizing guides for adult and children's clothing. UK/continental conversion sizes are shown. The company's philosophy page gives useful information on reasons for using natural products in clothing and in the home.
Links: Yes
On-line ordering: Yes
Credit cards: Visa; Mastercard
Email: mail@spiritofnature.co.uk
Address: Burrhart House, Cradock Road, Luton, Bedfordshire LU4 0JF
Tel: 0870 725 9885
Fax: 0870 725 9886

Splashdown Water Birth Services
www.waterbirth.co.uk

Specialities: Birth pools; water birth
Website: Pools for hire from local collection points throughout the UK. It is now possible for every mother-to-be to use a water birth pool at home or in hospital for NHS or private births anywhere in the UK. Website gives information on water births, pools, books and videos. There is full information on hiring pools, pool specifications and a booking form to print out.
Catalogue: Free information pack
Email: jayn@splashdown.org.uk
Address: 17 Wellington Terrace, Harrow-on-the-Hill, Middlesex HA1 3EP
Tel: 020 8422 9308
Fax: 020 8422 9308

Steiner Waldorf Schools Fellowship
www.steinerwaldorf.org.uk

Specialities: Schools; education; organisation
Website: The Fellowship represent 26 schools, 45 kindergartens and 8 teacher education seminars in the UK and Ireland. The website provides key perspectives, a listing of the schools with addresses and key information. There is also information on publications and teaching.
Email: mail@waldorf.compulink.co.uk
Address: Kidbrooke Park, Forest Row, East Sussex RH18 5JA
Tel: 01342 822 115
Fax: 01342 826 004

Straw Bale Building Association
www.strawbalebuildingassociation.org.uk

Specialities: Straw bale buildings; organisation
Website: The association is an informal grassroots group of people with an interest in straw bale building. There are irregular meetings and a newsletter. Website has headings on people who can help you build, help with drawings and design, help with building regulations, teaching workshops and useful organisations. There is also a list of other contacts.
Email: admin@strawbalebuildingassociation.org.uk
Address: Hollinroyd Farm, Butts Lane, Todmorden, West Yorkshire OL14 8RJ
Tel: 01706 818 126

Stringer Laboratories Ltd
www.neem-x.com

Speciality: Pest control
Website: Natural plant product derived from neem seeds which contain an effective repellent for controlling pests in organic farming and gardening systems. There is a printable order form on the website. Further information on neem may be had from www.neemfoundation.org in India.
Email: info@neem-x.com
Address: Sheraton House, Castle Park, Cambridge, Cambridgeshire CB3 0AX
Tel: 01223 370036
Fax: 01223 460178

Su Su Ma Ma World Wear
www.susumama.co.uk

Specialities: Baby clothing; children's clothing; fancy dress
Website: A small UK based company using fairtrade standards to produce a groovy rage of colourful clothing for kids sourced from around the world. On-line shop has headings for accessories, dresses, hats, jackets, knitwear, shirts, trousers, leggings and fancy dress.
On-line ordering: Yes
Credit cards: Most major credit cards
Email: info@susumama.co.uk
Address: 243 Holcroft Court, Clipstone Street, London, W1P 7DZ
Tel: 020 7436 6768
Fax: 020 7436 6768

Suffolk Herbs
www.suffolkherbs.com

Specialities: Seeds; vegetable seeds; herb seeds; wildflower seeds; gardening
Website: A very comprehensive seed site covering among others: herbs, herb lawns, wildflowers, oriental vegetables, salads, green manure, organic products and organic crop protection. This is a very comprehensive stockist of organic seed varieties
.**Links:** None
On-line ordering: Yes
Credit cards: Most major cards
Contact: Brian Haynes
Email: suffolkherbs@btinternet.com
Address: Monks Farm, Coggeshall Road, Kelvedon, Essex CO5 9PG
Tel: 01376 572456
Fax: 01376 571189

Suma Wholefoods
www.suma.co.uk

Speciality: Wholesaler
Website: A vegetarian worker's co-operative independent wholesaler distributing over 7,000 product lines. The website gives a fascinating history of the co-operative, its systems and its growth. There is a full on-line product index and information on own brand Suma lines. There are also recipes and job information.
Links: Yes
Contact: Rowena Ross
Email: info@suma.co.uk
Address: Unit AX1,
Dean Clough Industrial Estate, Halifax, West Yorkshire HX3 5AN
Tel: 01442 345513
Fax: 01422 349429

Sundance Market
www.sundancemarket.com

Specialities: Home delivery; fruit; vegetables; meat; groceries; bodycare
Website: Website provides on-line ordering of natural and organic food delivered throughout the UK. There is a useful navigation page to teach customers how to use the site.
On-line ordering: Yes
Credit cards: Visa; Mastercard, Delta; Switch
Email: admin@sundancemarket.com
Address: 250 Kings Road,
London SW3 5UE
Tel: 020 7376 3649
Fax: 0870 063 2432

SunDog (Renewable Energy & Environmental Services)
www.sundog-energy.co.uk

Specialities: Alternative energy; solar energy; wind energy
Website: Consultancy service providing access to a wide experience of renewable energy systems, particularly grid connected PV systems. Website gives information on system requirements, grid connected PV, wind turbines and system sizing. There are also photos and information on existing installations.
Email: info@sundog-energy.co.uk
Address: Fell Cottage, Matterdale End, Penrith, Cumbria CA11 0LF
Tel: 017684 82282
Fax: 017684 82600

Sundrum Organics
www.sundrum.co.uk

Specialities: Home delivery; vegetarian & vegan
Website: Home delivery of vegetarian and vegan wholefoods in West Central Scotland. Website was under construction at time of writing but there is an email facility for requesting a printed catalogue.
Catalogue: Yes
Contact: Steve Wall
Email: sales@sundrum.co.uk
Address: Unit 5,
High House Industrial Estate, Auchinleck, Strathclyde KA18 2A, Scotland
Tel: 01290 426770
Fax: 01290 426770

Super Globe
www.lightnet.co.uk

Specialities: Alternative technology; alternative energy
Website: Website gives information on alternative technology, free energy and environmental issues. The company sells relevant books.
Email: superglobe@lightnet.co.uk
Address: LightNet, PO Box 9640, London E11 2XY
Tel: 020 8518 8633
Fax: 020 8518 8633

Super Natural Ltd
www.commonwork.org

Specialities: Composts; liquid plant food; soil conditioner; gardening
Website: Organic composts and related products made from own cow manure and available through garden centres. Organic mulch, soil conditioners, composts and liquid plant foods, free from weed seeds. Website is a collective one for rural enterprises and educational charities at Bore Place in Kent.
Contact: Caroline Dunmall
Email: carolined@commonwork.org
Address: Bore Place, Chiddingstone, Edenbridge, Kent TN8 7AR
Tel: 01732 463255
Fax: 01732 440264

Supernature.org
www.supernature.org

Specialities: Bodycare; skincare; aromatherapy products; household; gifts
Website: High quality organic products at low prices. Range includes body and skincare, sun and haircare, aromatherapy, herbs and spices, gift sets and food, including organic chocolate. There is secure on-line ordering or orders can be faxed or emailed.
On-line ordering: Yes
Credit cards: Most major cards
Email: sales@supernature.org
Address: Freepost, Guernsey, Channel Islands GY1 5SS
Tel: 0845 458 2123
Fax: 0845 458 2133

Sussex High Weald Dairy Products
http://easyweb.easynet.co.uk/mhardy

Specialities: Dairy products; sheep's milk; cheese
Website: Dairy products made from organic sheep's milk and organic cow's milk. An order form may be downloaded from the website which gives details of the range of sheep's and cow's milk products. It also gives nutritional information, recipes and press cuttings.
Credit cards: None
Contact: Mark Hardy
Email: mhardy@agnet.co.uk
Address: Putlands Farm, Duddleswell, Uckfield, East Sussex TN22 3BJ
Tel: 01825 712647
Fax: 01825 712474

Sustain
www.sustainweb.org

Specialities: Organisation; farming; agriculture
Website: Campaigns for sustainable food and farming and advocates food and agriculture policies and practices that enhance the health and welfare of animals and people, improve the living and working environment, promote and enrich society and culture. Website provides news items and information on the organisation and its members and projects. Publications are listed and may be ordered on-line
Links: Yes
On-line ordering: Yes
Email: sustain@sustainweb.org
Address: 94 White Lion Street, London N1 9PF
Tel: 020 7837 1228
Fax: 020 7837 1141

Sustainable Travel and Tourism
www.sustravel.com

Specialities: Tourism; holidays
Website: Best practice on sustainable and environmentally sensitive travel and tourism. STT on-line provides a centralised source of information, updated on a regular basis with case studies, news and feature articles. It provides a medium for participating companies to provide news of their latest technological breakthroughs and success stories.
Email: sford@sustravel.com
Address: 14 Greville Street, London EC1N 8SB
Tel: 020 7871 0123
Fax: 020 7871 0111

Swaddles Green Farm
www.swaddles.co.uk

Specialities: Meat; poultry; ready-meals; children's food; charcuterie
Website: Certified with both the Soil Association and Organic Farmers and Growers. Home delivery of award winning products with fully personalised butchery service and on-line shopping facility. There is a soap box section with news, views and opinions and a cook book recipe section.
On-line ordering: Yes
Contact: Charlotte Reynolds
Email: information@swaddles.co.uk
Address: Hare Lane,
Buckland St Mary, Chard,
Somerset TA20 3ZB
Tel: 01460 234387
Fax: 01460 234591

The Swedish Window Company
www.swedishwindows.com

Specialities: Building materials; windows
Website: Windows and doors made from Scandinavian sustainable sources. Installation and disposal service. Website provides environmental information and articles from Greenpeace publications. Working sections and drawings can be downloaded following registration.
Links: Yes
Email: info@swedishwindows.com
Address: Millbank,
Earls Colne Industrial Park,
Earls Colne, Essex CO6 2NS
Tel: 01787 223931
Fax: 01787 220525

Take It From Here
www.takeitfromhere.co.uk

Specialities: Italian foods; olive oils; pasta; wines
Website: Specialises in quality Italian foods including organic olive oils and pasta. Also truffle products, pasta sauces, cheese, wines and vinegars. On-line ordering with detailed, illustrated and captioned catalogue. Gift boxes are available. (See also Danmar International).
Links: Yes
On-line ordering: Yes
Credit cards: Most major cards
Contact: Maureen & Pietro Pesce
Email: sales@tifh.co.uk
Address: Unit 04,
Beta Way,
Thorpe Industrial Estate,
Egham, Surrey TW2 8RZ
Tel: 0800 137064
Fax: 01784 477813

Tamarisk Farm
www.tamariskfarm.co.uk

Specialities: Meat; flours; vegetables; holidays; box scheme
Website: All produce home-grown on the family owned and run farm. Website provides details of produce available from the farm and which items are available by mail order. Other produce may be purchased at the farm itself. There is comprehensive information on the holiday cottages including a booking availability chart. There are also details of pedigree kittens for sale.
Credit cards: None
Contact: Adam Simon
Email: mail@tamariskfarm.co.uk
Address: West Bexington,
Dorchester, Dorset DT2 9DF
Tel: 01308 897781

Tamar Organics
www.tamarorganics.co.uk

Specialities: Seeds; fertilisers; composting; wildflowers; crop protection; gardening
Website: There is a comprehensive on-line catalogue with plant descriptions, sowing details and prices. A companion planting chart may be ordered. New organic gardening centre now open with everything for the organic gardener as well as organic food.
Links: None
On-line ordering: No
Credit cards: Visa; Mastercard; Switch
Export: Europe
Catalogue: Yes
Contact: C. Guilfoy
Email: tamarorganics@compuserve.com
Address: Tavistock Woodlands Estate, Gulworthy, Tavistock, Devon PL19 8DE
Tel: 01822 834887
Fax: 01822 834282

Taste Connection
www.tasteconnection.com

Specialities: Retail shop; meats; vegetables; dairy; grocery; olive oils
Website: Speciality and organic food shop with large deli counter and huge cheese selection. Specialises in olive oils from around the world. Website was under construction at the time of writing.
Contact: Iseut & Adrian Richards
Email: tasteconnection@aol.com
Address: 76 Bridge Street, Ramsbottom, Bury, Lancashire BL0 9AG
Tel: 01706 822175
Fax: 01706 822941

Taylors of Harrogate
www.feelgoodcoffee.co.uk

Specialities: Coffees; teas
Website: Organic coffees and teas. A company that takes its ethical responsibilities very seriously and builds long term relationships with its growers. Website provides information on where the coffees and teas originate, information on the products and how they can be obtained.
Email: sales_admin@bettysandtaylors.co.uk
Address: Pagoda House, Plumpton Park, Harrogate, North Yorkshire HG2 7LD
Tel: 05000 418898

The Tea and Coffee Plant
www.coffee.uk.com

Specialities: Coffees; teas
Website: Holders of an organic certificate for coffee roasting and tea blending. Wide range of coffees, roasts, teas, blends and herb teas. Certified organic, fairly traded and other fine produce. Retail, mail order and wholesale. The website provides pages on the history of coffee, roasting, all about tea, growing coffee and drinking coffee.
Links: Yes
Catalogue: Price list
Contact: Ian Henshall
Email: coffee@pro-net.co.uk
Address: 170 Portobello Road, London W11 2EB
Tel: 020 7221 8137

Teddington Cheese
www.teddingtoncheese.co.uk

Speciality: Cheese
Website: Cheeses from all over Britain and Europe produced on small farms using traditional methods. Website has headings for cheeses, cheese selections, hampers, other goods, gift store and wines. Over 130 cheeses are sold and they are classified on the site by country of provenance. Photographs, prices, full descriptions, tasting notes and map showing their origin are provided.
On-line ordering: Yes
Credit cards: All major credit and debit cards
Email: dougandtony@teddingtoncheese.co.uk
Address: 42 Station Road, Teddington TW11 9AA
Tel: 020 8977 6868

Terre De Semences Organic Seeds
www.terredesemences.com

Specialities: Seeds; vegetable seeds; flower seeds; gardening
Website: Organically or bio-dynamically raised vegetable and flower seeds. With over 1,400 varieties this is a major supplier of organically grown seed. The catalogue contains advice on how to save your own seed. A sample of the comprehensive catalogue entries is shown on the site. Over 1,000 organic seed varieties are available for purchase on-line. The full printed catalogue may also be ordered on-line.
On-line ordering: Yes
Catalogue: Yes (£5.00 inc p&p)
Contact: Chris Baur
Email: info@terredesemences.com
Address: Ripple Farm, Crundale, Nr Canterbury, Kent CT4 7EB
Tel: 01227 731815
Fax: 01227 767187

Tesco
www.tesco.co.uk

Specialities: Supermarket; home delivery; groceries; wines; baby foods
Website: Home delivery grocery service with on-line ordering for many areas. Special organic pages including recipes. Organic department is listed separately for on-line shopping.
On-line ordering: Yes
Credit cards: Most major cards
Email: online@tesco.co.uk
Tel: 0845 7225533

Textures
www.textilesfromnature.com

Specialities: Fabrics; herbal pillows; bedlinen; clothing; towels; paints
Website: Organic and eco-friendly clothing fabric and household goods. Range covers sisal flooring, organic unbleached linens, hemps wools, cottons and Auro Organic paints. Site includes information pages on cotton and hemp. On-line swatches can be seen on the site.
Catalogue: Fabric swatches available
Email: jag@textilesfromnature.com
Address:
84 Stoke Newington Church Street, London N16 0AP
Tel: 020 7241 0990
Fax: 020 7241 1991

Thames Fruit Ltd
www.kentnet.co.uk/thamesuk

Specialities: Fruit; vegetables; wholesaler
Website: Specialises in citrus, melons, soft fruit, peaches, nectarines, broccoli, garlic, lettuce and capsicums from Spain. Sales are to supermarkets, specialist wholesalers and food processors. Website gives general details of the company in English and Spanish.
Contact: Pepe Morant
Email: thamesuk@kentnet.co.uk
Address: Station Yard, Hop Pocket Lane, Paddock Wood, Tonbridge, Kent TN12 6DQ
Tel: 01892 834379
Fax: 01892 834379

Think Natural Ltd
www.thinknatural.com

Specialities: Natural health products; bodycare
Website: A major website devoted to natural health and bodycare. Thousands of natural health products are stocked. Website provides on-line ordering for the product range and also a huge amount of information including encyclopaedia content from Dorling Kindersley and contributions from expert journalists and natural health practitioners on specific conditions.
On-line ordering: Yes
Credit cards: Most major credit cards
Email: info@thinknatural.com
Address: Unit 7, Riverpark, Billet Lane, Berkhamsted, Hertfordshire HP4 1DP
Tel: 01442 299200
Fax: 01442 866977

Thompson & Capper Ltd
www.tablets2buy.com

Specialities: Contract manufacturers; vitamins; food supplements
Website: A complete service for the contract manufacture of private label mineral and vitamin supplements. There is a company profile on the website, together with details of facilities, manufacturing services, packaging options and other tablet manufacturing services.
Contact: Bill Whittaker
Email: enquiries@tablets2buy.com
Address: 9 - 11 Hardwick Road, Astmoor, Runcorn, Cheshire WA7 1PH
Tel: 01928 573734
Fax: 01928 580694

Thorogoods (Organic Meat) Specialists
www.thorogoodsorganicincorporated.co.uk

Specialities: Meat; poultry; allergies
Website: Small, private, family-run business with mail order. Produces home-made sausages using good quality organic pork, organic bacon and home cooked ham. Sausages can be made to order for allergy sufferers. Website gives product listing with prices and details of free local delivery and charged delivery elsewhere. Orders can be telephoned or faxed.
Catalogue: Brochure
Contact: Paul Thorogood
Address: 113 Northfields Avenue, Ealing, London W13 9QR
Tel: 020 8567 0339
Fax: 020 8566 3033

The 3 Ridings Coppice Group
www.three-ridings.org.uk

Speciality: Coppice management
Website: Co-operative of independent coppice workers raising public awareness of the ecological benefits of reinstating coppice management to broad-leaved woodland. Website provides illustrated information on coppicing. Tuition in woodland crafts and management and hedge-laying is offered. Website also gives details of the individual members of the co-operative.
Email: info@three-ridings.org.uk
Address: Aske Gardens, Richmond, North Yorkshire DL10 5HJ
Tel: 01653 618892

Thorncroft Ltd
www.thorncroft.ltd.uk

Specialities: Cordials; soft drinks
Website: Traditional home made soft drinks using fresh natural ingredients which retain their flavours and herbal benefits. Products are available in major supermarkets, healthfood stores, independent retailers and some off-licenses. Website gives product information, recipes and company background. There is an email response form for information on local stockists.
Contact: Michael Haigh
Email: thorncroftsales@cs.com
Address: Durham Lane Industrial Park, Eaglescliffe, Stockton on Tees, Cleveland TS16 0PN
Tel: 01642 791792
Fax: 01642 791793

Thursday Cottage
www.thursday-cottage.com

Speciality: Jams
Website: Suppliers of wholesome marmalades, jams and curds to the healthfood trades. The range includes four hand-made organic marmalades and one organic jam. Website gives background of this small-scale, quality producer and details of the products. There is a search facility for UK stockists. There are also recipes, newsletter, and diary dates.
Contact: Hugh Corbin
Email: sales@thursday-cottage.com
Address: Carswell Farm, Uplyme, Lyme Regis, Dorset DT7 3XQ
Tel: 01297 445555
Fax: 01297 445095

Tisserand Aromatherapy
www.tisserand.com

Specialities: Aromatherapy products; bodycare; babycare
Website: Pure authentic pre-blended essential oils from around the world. Lotions for the face and body, baby products, hand and foot treatments, antiseptic gels, moist tissues, vaporisers and accessories. The website gives details of the company, distribution, products and training.
Contact: Tonya Harman
Email: info@tisserand.com
Address: Newtown Road, Hove, East Sussex BN3 7BA
Tel: 01273 325666
Fax: 01273 208444

Tods of Orkney
www.stockan-and-gardens.co.uk

Specialities: Oatcakes; shortbread; biscuits; vegetarian
Website: Manufacturers of a variety of bakery goods in the Orkney Islands. Most of the oatcake range is suitable for vegetarians, The products contain no nuts, preservatives or artificial flavourings. Website gives information on the oatcakes, shortbread and handmade biscuits. There is an order form for mail order service which can be emailed or printed.
Contact: James Stockan
Email: info@stockan-and-gardens.co.uk
Address: The Bakery, 18 Bridge Street, Kirkwall, Orkney KW15 1HR, Scotland
Tel: 01856 873165
Fax: 01856 873655

Tofupill
www.tofupill.co.uk

Specialities: Food supplements; menopause
Website: Tofupill is a natural soya-based supplement developed for menopausal women to provide enough phytoestrogen to help maintain hormonal balance. There is a printable information leaflet on the website. There is also a secure on-line ordering facility. Details of other on-line suppliers are provided.
On-line ordering: Yes
Address: PitRok, PO Box 1416, London W6 9WH
Tel: 020 8563 1120
Fax: 020 8563 9987

Total Organics
www.totalorganics.com

Specialities: Fruit; vegetables; meat; groceries; wines; bodycare
Website: Over 1,600 lines of certified organic foods. Dedicated to retailing only organic food. There is also a retail shop in Glasgow. Website was still under construction at the time of writing.
Contact: Michael Noble
Email: info@totalorganics.com
Address: 104 Holburn Street, Aberdeen, Grampian AB10 6BY, Scotland
Tel: 01224 593959
Fax: 01224 458928

The Totally Baby Shop.com
www.netbabe.co.uk

Specialities: Babyfood; babywear; baby products
Website: One-stop on-line baby shop. This is solely an internet site so it gives complete dedication to the internet customer. Each month new innovative products are selected and there are also special monthly offers. The site is divided into babywear, organic food, baby products and this month's specials.
On-line ordering: Yes
Credit cards: Most major cards
Email: enquiries@thetotalbabyshop.com

Tourism Concern
www.tourismconcern.org.uk

Specialities: Tourism; organisation; holidays
Website: The UK's leading resource for ethical and sustainable tourism raising awareness of tourism's impact on people and their environments. The website provides sections on news, magazine, being there, resources, campaigns, education, fairtrade, forums and boycott among others.
Email: info@tourismconcern.org.uk
Address: Stapleton House, 277-281 Holloway Road, London N7 8HN
Tel: 020 7753 3330
Fax: 020 7753 3331

Tracklements
www.tracklements.co.uk

Specialities: Chutneys; sauces; mustards; dressings
Website: Producer of quality chutneys, mustards and sauces. Website lists products, gives extensive recipe suggestions and the company history. Overseas distributors are listed on the site.
Links: Yes
Contact: Guy Tulburg
Email: info@tracklements.co.uk
Address: The Dairy Farm, Pinkney Park, Sherston, Malmesbury, Wiltshire SN16 0NX
Tel: 01666 840851
Fax: 01666 840022

Trafalgar Fisheries
www.trafish.com

Speciality: Trout
Website: One of the earliest commercial fish farms in the UK with fish stocked at low density in natural gravel ponds allowing them to swim freely and mature naturally. Brown trout and rainbow trout are certified organic. Site gives full details on all the products and processing as well as recipes. Trout are available for stocking.
Contact: John Williams
Email: info@trafish.com
Address: Barford Fish Farm, Downton, Wiltshire SP5 3QF
Tel: 01725 5106148
Fax: 01725 511165

Traidcraft plc
www.traidcraft.co.uk

Specialities: Organisation; honey; tea; chocolate.
Website: Motivated by Christian concern for fairtrade, this is an importer and distributor of organic honey, tea and coffee. The extensive website explains the reasons behind Traidcraft and has pages on wholesale, retail, fairtrade, and educational resources.
Contact: Joe Osman
Email: traidcraft@globalnet.co.uk
Address: Kingsway, Gateshead, Tyne and Wear NE11 0NE
Tel: 0191 491 0591
Fax: 0191 482 2690

Trannon
www.trannon.com

Speciality: Furniture
Website: Pioneers of sustainable design and clean production of furniture, using steam-bending to create elegant ergonomic curves whilst seasoning the timber quickly and using a fraction of the energy of conventional methods. Green ash thinnings from local woodlands are used as raw materials. Website gives information on a variety of pieces and their prices.
Contact: Roy Tam & Tony Minx
Email: info@trannon.com
Address: Chilhampton Farm, Wilton, Salisbury, Wiltshire SP2 0AB
Tel: 01722 744577
Fax: 01722 744477

Trayner Pinhole Glasses
www.trayner.co.uk

Speciality: Eyecare
Website: Special glasses developed from the traditional pinhole principle are helpful with short-sight, long-sight, presbyopia, headaches, eyestrain and computer work. Glasses are locally produced in the UK. The site explains the principles behind the glasses and gives information on the human eye, eye exercise and eye nutrition.
On-line ordering: Yes
Contact: Peter Duthie
Email: sales@trayner.co.uk
Address: 12 - 14 Old Mill Road, Portishead, Bristol BS20 7EG
Tel: 07967 180867
Fax: 01749 85083

Treske Ltd
www.treske.co.uk

Speciality: Furniture
Website: Extensive range of eco-friendly furniture using UK hardwoods grown under sustainable woodland husbandry and processed naturally. A full catalogue can be requested by email. Website illustrates furniture in room settings. It also sets out Treske's unique service and philosophy and provides a guide on how to choose furniture. There is information on the visitor centre and factory shop.
Catalogue: Yes
Email: treske@btinternet.com
Address: Station Maltings, Thirsk, Yorkshire YO7 4NY
Tel: 01845 522770
Fax: 01845 522692

Triodos Bank
www.triodos.co.uk

Specialities: Banking; financial
Website: One of Europe's leading ethical banks. Website explains the way it finances initiatives, delivering real social and environmental benefits and how saving and investing with Triodos can make a real difference. It provides information on the full range of services including saving and investing, personal accounts, partnership accounts and business and charity accounts.
Email: mail@triodos.co.uk
Address: Brunel House, 11 The Promenade, Clifton, Bristol BS8 3NN
Tel: 0117 973 9339
Fax: 0117 973 9303

Tropical Forest Products
www.tropicalforest.com

Speciality: Honey
Website: Distinctive honeys from throughout the tropics including Zambia, Vietnam, Cuba and the Solomon Islands. Zambian Organic Forest Honey is a fairly traded honey which is gathered from wild bees living in traditional hives deep in the forest. Website gives details of individual honeys with their origins and characteristics.
Contact: David Wainwright
Email: mail@tropicalforest.com
Address: PO Box 92, Aberystwyth, Ceredigion SY23 1AA Wales
Tel: 01970 832511
Fax: 01970 832511

Tropical Wholefoods
www.tropicalwholefoods.com

Speciality: Dried fruits
Website: Sun dried organic and natural tropical fruits from all over the world. The range includes pure fruit, fruit bars, tropical mueslis, mixes, exotic mushrooms, sun dried tomatoes and spices, all fairly traded. The range is listed on the website and there are recipe suggestions. There is also a reprint of a *Daily Telegraph* article on the company.
Contact: Kate Sebag
Email: info@tropicalwholefoods.com
Address: Unit 9, Industrial Estate Hamilton Road, London SE27 9SF
Tel: 020 8670 1114
Fax: 020 8670 1117

TUCANO
www.tucanobeach.com

Speciality: Clothing
Website: A range of beach and casual clothing for men women and children using natural and organic materials. Clothing is illustrated on the website with prices.
On-line ordering: Yes
Credit cards: Visa; Mastercard; Switch; Delta
Email: mail@tucagua.com
Address: Pound House, Pound Road, West Wittering, West Sussex PO20 8AJ
Tel: 01243 513 757
Fax: 01243 671 884

21ST CENTURY HEALTH
www.21stcenturyhealth.co.uk

Specialities: Bodycare; aromatherapy products; vitamins; nutritional supplements; household products
Website: Products with no animal testing for a chemical-free home, body and spirit. On-line catalogue with photographs, descriptions, prices and buying information.
On-line ordering: Yes
Credit cards: Most major credit cards
Catalogue: Yes
Email: info@21stcenturyhealth.co.uk
Address: 3 Water Gardens, Stanmore, Middlesex HA7 3QA
Tel: 020 7935 5440
Fax: 020 7487 3710

TY-MAWR LIME
www.lime.org.uk

Specialities: Lime products; paints; building materials
Website: Supplier of lime-based products: washes, putties, plaster pigments and also milk based paints and emulsions. Website gives details of products and services. Site was under construction at time of writing.
Links: Yes
Email: tymawr@lime.org.uk
Address: Ty-Mawr Farm, Llangasty, Brecon, Powys LD3 7P, Wales
Tel: 01874 658249
Fax: 01874 658502

UK SOCIAL INVESTMENT FORUM
www.uksif.org

Specialities: Organisation; finance; ethical investment
Website: Provides a listing of members in the field of ethical investments including unit trust and independent financial advisers. The website gives an indexed directory of members and affiliates. There is also a library with documents listed by subject and document type. Details are given of the programme of activities undertaken by the organisation.
Email: info@uksif.org
Address: Holywell Centre, 1 Phipps Street, London EC2A 4PS
Tel: 020 7749 4880
Fax: 020 7749 4881

UNICORN GROCERY
www.unicorn-grocery.co.uk

Specialities: Groceries; wines; gluten-free products; sugar-free products
Website: Large wholesome foodstore specialising in organic and fresh produce. A worker co-operative owned and run by its workforce. Website provides a complete downloadable price list. There are also sections on ethics, organics, jobs and recipes.
Contact: Kellie Bubble
Email: office@unicorn-grocery.co.uk
Address: 89 Albany Road, Chorlton, Manchester M21 0BN
Tel: 0161 861 0010
Fax: 0161 861 7675

Unit Energy
www.creationday8.co.uk

Speciality: Alternative energy
Website: Independent domestic supplier of renewable electricity. Website explains what makes it special, how it works and what it costs. Registration may be made on-line or alternatively documents can be downloaded and a contract may be filled in.
Email: enquiries@unit-energy.co.uk
Address: Unit E House,
16 Avon Reach, Monkton Hill,
Chippenham,
Wiltshire SN15 1EE
Tel: 01249 705550
Fax: 01249 445374

Urban Organics
www.urbanorganics.co.uk

Specialities: Retail shop; box scheme
Website: Delivery scheme in Cardiff. Orders can be placed by email, phone or fax. Website covers the shop, the box scheme, current week's produce and recipes.
Credit cards: Most major credit cards
Email: sales@urbanorganics.co.uk
Address: 32 Splott Road, Splott,
Cardiff, Glamorgan CF24 2DA, Wales
Tel: 029 2040 3399

H Uren & Sons Ltd
www.uren.com

Specialities: Food industry supplier; processors; repackers; distributors
Website: Organic ingredients for the food industry. Organic sorting, inspection and packing. Website gives details of the company and its products - quick frozen fruits, fruit purées and fruit juice concentrates. There is a page detailing organic product lines.
Email: james.uren@uren.co.uk
Address: Woodpark, Neston,
South Wirral, Cheshire CH64 7TB
Tel: 0151 353 0330
Fax: 0151 353 0251

Utobeer
www.utobeer.co.uk

Speciality: Beer
Website: Organic beers. A specialist company dealing in all aspects of beer from selling bottled beers, through food produced from beer to home brewing as well as a resource offering advice and assistance. The website was under construction at the time of writing.
Contact: Richard Dinwoodie
Email: info@utobeer.co.uk
Address: PO Box 30053,
London SE1 6XT
Tel: 0870 901 2337
Fax: 0870 901 2338

UVO (UK) Ltd
www.grander.com

Speciality: Water purifiers
Website: Distributor of devices which revitalise water and restore its primeval force, regenerating itself and improving its capacity to purify itself. The principles are based on those of the Austrian naturalist Johan Grander. Website gives details of the various products including an air revitaliser and information on books and videos.
Email: uvo.uk@virgin.net
Address: PO Box 160, Hay on Wye,
Herefordshire HR3 6ES
Tel: 01497 831 029
Fax: 01497 831 420

Vaccination Awareness network UK
www.van.org.uk

Specialities: Health information; organisation; childcare
Website: Provides unbiased information on vaccinations. Non-profit making, volunteer staffed organisation which provides information on meningitis c, mmr, dpt, polio, hib, tetanus, flu, hepatitis b, tb and others on the website. Also gives membership details.
Contact: Joanna Jarpasea-Jones
Email: enquiries@van.org.uk
Address: 159 Sneinton Dale, Nottingham, Nottinghamshire NG2 4LW
Tel: 0870 444 0894
Fax: 0870 741 8415

Valuepharm Ltd
www.valuepharm.ie

Specialities: Food supplements; herbal remedies
Website: Website was under construction at time of writing.
Contact: John O'Callaghan
Email: valuepharm@eircom.net
Address: Market Square, Carlingford, Co Louth, Ireland
Tel: + 353 429 37397
Fax: + 353 429 373344

Vegan Organic Network
www.veganvillage.co.uk

Specialities: Vegan; organisation; directory; resource
Website: A network of vegan contacts and organisations. The site covers vegan shopping with on-line shops, vegan shoes and cosmetics; vegan food and drink with recipes, restaurants and UK suppliers; vegan social with homepages, contacts, vegan groups and clubs; vegan travel with UK guest houses, travel guides and links abroad; vegan health with nutrition links and campaigning groups; and vegan business with accountants, solicitors and graphic designers. There is also a noticeboard, a site of the week and what's new.
Email: postie@veganvillage.co.uk
Address: 58 High Lane, Chorlton, Manchester M21 9DZ
Tel: 0161 860 4869
Fax: 0161 860 4869

Vegan Society
www.vegansociety.com

Specialities: Vegan; organisation
Website: Society promoting vegan lifestyles or ways of living that seek to exclude all forms of exploitation of animals for food, clothing or any other purpose. The society promotes through books, videos and publications, through a magazine, through a trademark, through contact groups and through information. Website gives details of the society, information sheets, publications and products. There is an FAQ page.
Links: Yes
Email: info@vegansociety.com
Address: Donald Watson House, 7 Battle Road, St Leonards on Sea, East Sussex TN37 7AA
Tel: 01424 427 393
Fax: 01424 717 064

Veganline
www.animal.nu

Specialities: Footwear; vegan
Website: Veganline stocks vegan shoes, belts, hemp shoes, bags, sneakers and T shirts. UK postage is included in the prices. Descriptive text is available in a number of languages.
Links: Yes
On-line ordering: Yes
Credit cards: Most major credit cards
Email: veg@animal.nu
Address: Freepost LON10506, London SW14 1YY
Tel: 0800 458 4442
Fax: 020 8878 3006

Veganstore
www.veganstore.co.uk

Specialities: Footwear; clothing; groceries; bodycare; healthcare; vegan
Website: One-stop vegan shop with over 300 animal and cruelty-free products. Website covers cruelty-free footwear, fake jackets, bodycare and cosmetics, vitamins and healthcare, groceries, sweet things, candles, books and clothing. Company regularly attends vegan festivals and will notify customers who subscribe to the newsletter. Orders may be placed on-line, by phone and by fax. There is no retail outlet as this is a mail order company only.
On-line ordering: Yes
Credit cards: Most major credit cards
Email: info@veganstore.co.uk
Address: 15 Chichester Drive East, Salt Dean, Brighton, East Sussex BN2 8LD
Tel: 01273 302 979
Fax: 01273 302 979

Vegetarian Cookery Courses
www.ashburtoncentre.co.uk

Specialities: Cookery courses; vegetarian; holidays
Website: The centre offers a chance to participate in the creation of a real sense of community, learning new skills and making new friends. The website provides information on the school of cookery and the courses which include organic vegetarian, Mediterranean, Italian, chalet and boat crew cookery. The school uses organic and natural ingredients. Holidays, courses and retreats are held in Devon and Spain.
Catalogue: Brochure
Email: stella@ashburtoncentre.co.uk
Address: The Ashburton Centre, 79 East Street, Ashburton, Devon TQ13 7AL
Tel: 01364 652 784
Fax: 01364 653 825

Vegetarian Pages
www.veg.org

Specialities: Resource; vegetarian
Website: Aims to be the premier website for vegetarian information. Website was still under re-construction at the time of writing but had a link to the Old Vegetarian pages.

Vegetarian Society
www.vegsoc.org

Specialities: Organisation; vegetarian
Website: Represents the interests of the UK's 4 million vegetarians by informing, educating, campaigning and working with food and health professionals. Membership form can be printed. Extensive site with membership information, history, news, membership discounts and aims.
Contact: Vanessa Brown
Email: info@vegsoc.org
Address: Park Dales, Dunham Road, Altrincham, Cheshire WA14 4QG
Tel: 0161 925 2000
Fax: 0161 926 9182

Veggi Wash Food Safe
www.food-safe.com

Specialities: Household products; healthcare; food hygiene
Website: The company provides a range of unique formulations to improve the health and safety of the family and the food preparation and processing industry by improving the health and safety of the food we eat and the area in which we prepare and eat it. Website provides pages on the concept, the products, industry solutions and consultancy, certification and ordering. Products are available through healthfood outlets but may also be ordered direct.
Contact: Bruce Green
Email: bruce.green@pipemedia.co.uk
Address: Food Safe Ltd, Winwick, West Haddon, Northampton, Northamptonshire NN6 7PD
Tel: 01788 510415
Fax: 01788 510515

Vegi Ventures
www.vegiventures.com

Specialities: Holidays; vegetarian & vegan
Website: Holidays in Britain, Turkey and Peru with vegetarian and vegan food, small friendly groups, and environment-oriented approach. Website gives details of the company, the holidays with dates, the food and specials. There is a comments page and a newsletter. A printed brochure can be requested on an email form.
Links: Yes
Catalogue: Brochure
Contact: Nigel & Jacky Walker
Email: holidays@vegiventures.com
Address: Castle Cottage, Castle Acre, Norfolk PE32 2AJ
Tel: 01760 755 888
Fax: 01760 755 888

The Village Bakery
www.village-bakery.com

Specialities: Bread; jams; flour; vegetarian & vegan; restaurant; cookery courses
Website: Specialist bakers of organic breads and cakes in traditional wood fired bread ovens. Many foods are geared for people on special diets. There are bread making courses. The restaurant uses only organic ingredients and serves organic wines and beers. The website gives details of mail order, the restaurant, wholesale, stockists, the bakeshop and courses in bread making. Products are coded for special diets and vegetarians and vegans and may be ordered on-line.
Links: Yes
On-line ordering: Yes
Export: Yes
Catalogue: Brochures
Contact: Andrew Whitley
Email: info@village-bakery.com
Address: Melmerby, Penrith, Cumbria CA10 1HE
Tel: 01768 881515
Fax: 01768 881848

Vinceremos Wines & Spirits
www.vinceremos.co.uk

Specialities: Wines; spirits; beers; vegetarian & vegan; cordials; olive oil
Website: Runs the HDRA Wine Club. Stocks both European and New World wines. Dryness and fullness ratings are provided. A printed catalogue may be ordered on-line. There are 300 organic wines, beers, juices, cordials and spirits to choose from and wines are listed by country of origin. There is a search facility, including by price range. Company can supply information on wines acceptable to vegans and vegetarians.
On-line ordering: Yes
Catalogue: Yes
Contact: Harriet Walsh
Email: info@vinceremos.co.uk
Address: 19 New Street, Horsforth, Leeds, West Yorkshire LS18 4BH
Tel: 0113 205 45445
Fax: 0113 205 4546

Vintage Roots
www.vintageroots.co.uk

Specialities: Wines; beers; ciders; juices; vegetarian & vegan
Website: Possibly the largest shipper of organic wines and beers. Website gives information pages on vegetarian and vegan, bio-dynamic, GMOs and reconversion. Full ordering information provided.
On-line ordering: Yes
Catalogue: Yes (gratis)
Contact: Neil Palmer
Email: info@vintageroots.co.uk
Address: Farley Farms, Bridge Farm, Arborfield, Berkshire RG2 9HT
Tel: 0118 976 1999
Fax: 0118 976 1998

Viridian Nutrition Ltd
www.viridian-nutrition.com

Specialities: Vitamins & minerals; food supplements; herbal remedies
Website: A new company dedicated to producing an exceptional range of vitamins, minerals, and herbs and to making a significant contribution to the funds of environmental, children's and other selected charities. A minimum of 50% net profit is donated annually. Website gives pages on philosophy, purity, charity, animal rights, environment, products and store locator.
Links: Yes
Contact: Cheryl Thallon
Email: cheryl@viridian-nutrition.com
Address: 31 Alvis Way,
Royal Oak, Daventry,
Northamptonshire NN11 5PG
Tel: 01327 878050
Fax: 01327 878335

Vital Touch
www.vitaltouch.com

Speciality: Aromatherapy products
Website: Vital Touch brings together knowledge and experience of therapeutic massage and the ancient healing arts of reflexology and aromatherapy. The website provides a brief introduction to aromatherapy and gives a listing of products available for on-line ordering. These include items for pregnancy, birth and new parents, full home aromatherapy kits and bespoke products. There is a gift wrapping service.
Links: Yes
On-line ordering: Yes
Credit cards: Access; Visa; Mastercard; Solo; Switch
Contact: Katie Whitehouse
Address: PO Box 108, Totnes,
Devon TQ9 5UZ
Tel: 01803 840670
Fax: 01803 840113

Vitrition Ltd
www.optivite.co.uk

Speciality: Animal feeds
Website: Organic animal feeds and related products with nationwide delivery service. Website gives details of products.
Contact: Paul Forster
Email: optivite@cs.com
Address: Main Street,
Laneham, Retford,
Nottinghamshire DN22 0NA
Tel: 01777 228741
Fax: 01777 228737

Viva
www.viva.org.uk

Specialities: Organisation; magazine; vegetarian & vegan; wine; chocolate
Website: Viva is a vegetarian and vegan charity which produces a quarterly magazine and provides information to people wanting to change their diets. It also campaigns to end factory farming of animals. The website provides campaign information, press releases and joining information. There is also an on-line shop selling vegan wine, beer and spirits, chocolates, gifts and stationery.
On-line ordering: Yes
Contact: Jo Lacey
Email: info@viva.org.uk
Address: 12 Queen Square,
Brighton, East Sussex BN1 3FD
Tel: 01273 777688
Fax: 01273 776755

VMM (Vegetarian Matchmakers)
www.veggiematchmakers.com

Specialities: Introduction agency; vegetarian & vegan
Website: The leading dating agency exclusively for vegetarians and vegans. The website provides information on how the agency works, member profiles and success stories. There is an FAQ page and an on-line enrolment form.
Links: Yes
Email: vmm@cybervillage.co.uk
Address: Concord House, 7 Waterbridge Court, Appleton, Warrington, Cheshire WA4 3BJ
Tel: 01925 601 609
Fax: 01925 860 442

Wafcol
www.wafcol.co.uk

Speciality: Pet food
Website: Producers of organic cat and dog foods and a range of exclusion diets. The website has section headings on nutritional information, problem solving, and sample requests as well as a product listing.
Address: Haigh Avenue, Stockport, Cheshire SK4 1NU
Tel: 0161 480 2781
Fax: 0161 474 1896

Waitrose
www.waitrose.com

Specialities: Supermarket; groceries; home delivery; wines; baby food
Website: Organics Direct is the on-line organic delivery service. Nearly 1,100 organic lines are stocked. A listing of organic products other than fruit and vegetables is provided on-line. There is an on-line delivery service for organic fruit, vegetables and salads anywhere on the UK mainland. There are also various information pages on organics available on the site.
On-line ordering: Yes
Credit cards: Most major cards

Walter Segal Self Build Trust
www.segalselfbuild.co.uk

Specialities: Architectural design; self-build; courses
Website: Company will advise on their environmentally friendly approach to building and will suggest architects who have the appropriate experience. An information pack is available. Website gives wide ranging information including examples of what others have built, publications, discussion and community buildings. Details of very affordable self-build courses are given.
Links: Yes
Catalogue: Information pack
Contact: Gillian Simmons
Email: info@segalselfbuild.co.uk
Address: 15 High Street, Belford, Northumberland NE70 7NG
Tel: 01668 213 544
Fax: 01668 219 247

Waste Watch
www.wastewatch.org.uk

Specialities: Recycling; organisation
Website: Six monthly guide to recycled materials, manufacturers and suppliers. Covers garden furniture, soil and pots as well as clothing, paper and stationery. The website provides extensive information on its activities and aims, on projects, schools and kids and its policy.
Links: Yes
Contact: Barbara Herridge
Email: info@wastewatch.org.uk
Address: Europa House, 13 - 17 Ironmonger Row, London EC1V 3QG
Tel: 020 7253 6266
Fax: 020 7253 5962

Water Dynamics
www.waterdynamics.co.uk

Specialities: Water saving; recycling
Website: Recycling systems for rain water and domestic grey water for re-use in toilet flushing and gardens. Up to 40% of water consumption can be saved. Website gives overall information with an email request form to apply for more details.
Contact: Information brochure
Email: info@waterdynamics.co.uk
Address: Unit 32, Branbridges Industrial Estate, East Peckham, Kent TN1 5HF
Tel: 01622 873322
Fax: 01622 873399

Waveney Rush Industry
www.waveneyrush.co.uk

Specialities: Flooring; rush-weave
Website: Waveney Rush revived the craft in 1947 and makes natural flooring with hand-made rush-weave floor coverings. It also makes log baskets, pet baskets, waste-paper baskets and chair seats. All products are made to order. Website gives products listing with information on how they are made.
On-line ordering: Yes
Credit cards: Mastercard; Visa; Switch; Delta; Amex
Email: crafts@waveneyrush.co.uk
Address: The Old Maltings, Caldecott Road, Oulton Broad, Lowestoft, Suffolk NR32 3PH
Tel: 01502 538777
Fax: 01502 538477

Weleda
www.weleda.co.uk

Specialities: Toiletries; babycare; homoeopathic remedies; herbal remedies; anthroposophic medicines
Website: Organic and bio-dynamic toiletries for adults and babies, available in healthfood shops and by mail order direct. Bio-dynamic herbal and homoeopathic remedies for all types of ailments. There is on-line ordering on the extensive website which also includes a remedy finder by clicking on the relevant illness. There is also an email facility for health advice from a pharmacist and for queries and special orders.
On-line ordering: Yes
Address: Heanor Road, Ilkeston, Derbyshire DE7 8DR
Tel: 0115 944 8200
Fax: 0115 944 8210

Well4ever
www.well4ever.com

Speciality: Food supplements
Website: A range of nutritional supplements with a specific aim - to produce effective solution-oriented healthcare programmes. Website was under construction at time of writing.
Contact: Robert Brydges
Email: rob@well4ever.com
Address: Wellness Holdings, 7 - 11 Kensington High Street, London W8 5NP
Tel: 020 7368 3345
Fax: 020 7368 3346

Welsh Fruit Stocks
www.welshfruitstocks.co.uk

Specialities: Garden nursery; soft fruit plants; gardening
Website: Propagators of top quality soft fruit plants grown in isolated and extremely healthy conditions on increasingly organic principles. Plants can be collected in season or sent by carrier. Supplies both gardeners and commercial enterprises. Available fruit varieties are catalogued on the website.
Links: None
Email: sian@welshfruitstocks.co.uk
Address: Bryngwyn, Kington, Hereford, Herefordshire HR5 3QZ
Tel: 01497 851209
Fax: 01497 851209

Welsh Hook Meat Centre
www.welsh-organic-meat.co.uk

Specialities: Meat; poultry; game; wholesalers
Website: Extensive mail order list. Supplies meat from Soil Association certified farms. Also supplies the catering industry with organic products. Orders may be placed by phone, fax, post or email.
Credit cards: Visa; Mastercard
Email: welshhookmeat@talk21.com
Address: Woodfields,
Withybush Road, Haverfordwest,
Pembrokeshire SA62 4BW, Wales
Tel: 01437 768876
Fax: 01437 768877

Welsh Institute of Organic Studies
www.wirs.aber.ac.uk/research /organic

Specialities: Research; courses; degrees;
Website: Conducts research and development of organic farming systems and offers a BSc course in agriculture with organic agriculture. Training courses are geared to producers.
Contact: Dr N H Lampkin
Email: nhl@aber.ac.uk
Address: University of Wales,
Aberystwyth,
Ceredigion SY23 3AL, Wales
Tel: 01970 622248
Fax: 01970 622238

West Usk Lighthouse
www.westusklighthouse.co.uk

Specialities: Bed & breakfast; holidays; flotation tank
Website: Website provides photographs, tariff and descriptions of this unusual guest house. Complementary therapies are available and include aromatherapy, reflexology, past life regression, kinesiology, reiki and shiatsu among others. A flotation tank is available.
Email: lighthouse@tesco.net
Address: Lighthouse Road,
St Brides, Wentloog, Newport,
Gwent NP10 8SF, Wales
Tel: 01633 810 126
Fax: 01633 815 860

Westcountry Organics Ltd
www.westcountryorganics.co.uk

Specialities: Meat; cheese; box scheme; home delivery
Website: Organic food and drink delivered to the door nationwide by overnight by courier. Everything from meat and vegetables to wines and butter. Orders can be placed on-line or by email, phone, fax or post. There is a downloadable order form on the website.
Links: Yes
On-line ordering: Yes
Credit cards: Most major credit cards
Catalogue: Yes
Email:
enquiries@westcountryorganics.co.uk
Address: Natson Farm,
Tedburn St Mary, Exeter,
Devon EX6 6ET
Tel: 0164 724 724
Fax: 0164 724 031

Westons
www.westons-cider.co.uk

Specialities: Cider: soft drinks; spritzer; vegetarian & vegan
Website: Brand and market leading organic cider and organic spritzer. Also an organic non-alcoholic drink, a blend of spring water, apple juice, rosewater, camomile, St Johns Wort, nettle and echinacea. Suitable for vegetarians and vegans. There is a guide to cider making on the website as well as a history of Westons, a product listing and on-line ordering.
On-line ordering: Yes
Credit cards: Most major credit cards
Contact: Roger Jackson
Email: tradition@westons-cider.co.uk
Address: H Weston & Sons Ltd,
The Bounds, Much Marcle, Ledbury,
Herefordshire HR8 2NQ
Tel: 01531 660233
Fax: 01531 660619

What Doctor's Don't Tell You
www.wddty.co.uk

Specialities: Health information; resource
Website: Monthly newsletter providing information on health issues which apparently the medical establishment would rather conceal. 100 or so back issues covering virtually every illness, drug and medical procedure plus an impressive library of booklets on specific conditions and how they may be treated can be ordered on-line from this site. There is also a facility for on-line subscriptions to the monthly newsletter.
Links: Yes
On-line ordering: Yes
Address: 2 Salisbury Road, London SW19 4YY
Tel: 0800 146 054

Whole Earth Foods Ltd
www.wholeearthfoods.co.uk

Specialities: Cereals; canned goods; soft drinks
Website: Pioneering organic food company makes a wide range of organic foods including breakfast cereals, peanut butter, soft drinks (including organic cola), baked beans, jams and coffee substitutes. See separate entry for Green and Black chocolate. Supports farmers worldwide who do not use pesticides, herbicides or chemical fertilisers. At time of writing website gives general background company information whilst under construction and offers an email facility to advise when it goes live.
Contact: Cluny Brown
Email: enquiries@wholeearthfoods.co.uk
Address: 2 Valentine Place, London SE1 8QH
Tel: 020 7633 5900
Fax: 020 7633 5901

Wiggly Wigglers Ltd
www.wigglywigglers.co.uk

Specialities: Composting; earthworms; books; gardening
Website: Earthworms and compost systems for composting. Also supplies books on composting. Offers advice on gardening and composting with worms and all aspects of conservation and recycling. On-line ordering facility.
Links: Yes
On-line ordering: Yes
Credit cards: Visa; Switch; Mastercard; Delta
Catalogue: Brochure
Contact: Heather Gorringe
Email: wiggly@wigglywigglers.co.uk
Address: Lower Blakemere Farm, Blakemere, Herefordshire HR2 9PX
Tel: 01981 500 391
Fax: 01981 500 108

Wigham Farm Ltd
www.wigham.co.uk

Specialities: Bed & breakfast; meat; poultry; dairy; holidays
Website: Organic farm and guest house. Organic products include table birds, lamb, beef, pork and pork products, and dairy produce. The website gives details of bed and breakfast with photographs of rooms and room rates. Availability can be checked by email.
Credit cards: Visa; Mastercard; Switch; Delta; Amex
Contact: Steve Chilcott
Email: info@wigham.co.uk
Address: Wigham, Morchard Bishop, Devon EX17 6RJ
Tel: 01363 877350
Fax: 01363 877350

Wild and Free Travel
www.dolphinswims.co.uk

Specialities: Dolphins; holidays; activity holidays
Website: A unique opportunity to swim with wild, free, friendly dolphins in the Red Sea. Website provides information on the dolphins, the location and local attractions. There are also accommodation details, special needs, dates, prices, booking information, booking form, personal stories, recent articles, dolphin books and photo gallery among the other sections on the website.
Links: Yes
Email: wild.dolphin@ntlworld.com
Address: 6 Old Bridge Court, Forres, Moray IV36 1ZR, Scotland
Tel: 01309 671 726

Wild Oceans Holidays
www.wildwings.co.uk

Specialities: Activity holidays; holidays
Website: Wildwings offer environment-caring activity holidays worldwide including bird watching, wild oceans, deep-sea exploration and expedition cruises. Full details, dates and prices of the holidays are available on the website. There is also a connection with a Dutch based tour operator operating working conservation holidays around the world under the name Ecovolunteer.
Credit cards: Most major credit cards
Email: wildinfo@wildwings.co.uk
Address: 1st Floor,
577-579 Fishponds Pond,
Bristol, BS16 3AF
Tel: 0117 965 8333

Wild Ginger Vegetarian Bistro
www.veganvillage.co.uk /wildginger

Specialities: Vegetarian foods; restaurant; vegetarian & vegan
Website: 100% vegetarian foods in this licensed restaurant. Also caters for special and exclusion diets. Website gives menus, events and press comments.
Links: Yes
Contact: Rachel Melton
Email: wildginger@veganvillage.co.uk
Address: 5 Station Parade, Harrogate, North Yorkshire HG1 1UF
Tel: 01423 566122
Fax: 01423 520056

The Wildlife Trusts
www.wildlifetrusts.org

Specialities: Organisation; wildlife; conservation
Website: Leading conservation charity concerned solely with wildlife with over 300,000 members. It is a partnership of 46 Wildlife Trusts which administer more than 2,000 nature reserves throughout the UK to protect wildlife for the future. The website has background information on the trusts and their work, information on how to take part and how to join and support them.
Email: info@wildlife-trusts.cix.co.uk
Address: The Kiln Waterside,
Mather Road, Newark,
Nottinghamshire NG24 1WT
Tel: 01636 677711
Fax: 01636 670001

Wilkinet Baby Carrier
www.wilkinet.co.uk

Speciality: Baby carriers
Website: Website provides reasons for using a baby carrier, how they can help, and how to use them. The site shows the products and accessories with prices and colour availability. Orders can be placed by email, phone, fax or post.
Credit cards: Visa; Mastercard; Switch
Contact: Sally Wilkins
Email: wilkinet.web@ntlworld.com
Address: PO Box 20,
Cardigan SA43 1JB, Wales
Tel: 01239 841844
Fax: 01239 841390

Willing Workers on Organic Farms
www.phdcc.com/wwoof

Specialities: Organisation; farming; agriculture; holidays; activity holidays
Website: A worldwide exchange network where bed and board and practical experience are given in return for work on organic farms for holiday breaks, travel or changing to a rural life. The aims of WWOOF are set out and there are pages of background information. Membership information is available on the site. International contact details are given for overseas WWOOF organisations.
Links: Yes
Email:
fran@wwoof-uk.freeserve.co.uk
Address: PO Box 2675, Lewes, East Sussex BN7 1RB
Tel: 01273 476286
Fax: 01273 476286

Wind and Sun Ltd
www.windandsun.co.uk

Specialities: Energy; solar energy; wind energy
Website: Supplier and installer of domestic natural and renewable energy systems to the customer's requirements. The website provides comprehensive information on products and services including on-line ordering from the catalogue. It gives details of projects, its workshop and offices, demonstration sites and estimating requirements. There are also special offers.
Links: Yes
On-line ordering: Yes
Credit cards: Most major credit cards.
Catalogue: Yes
Email: info@windandsun.co.uk
Address: Humber Marsh, Stoke Prior, Leominster, Herefordshire HR6 0NE
Tel: 01568 760671
Fax: 01568 760484

Windmill Hill City Farm Shop
www.windmillhillcityfarmshop .org.uk

Specialities: Meat; poultry; eggs; vegetables; dairy; café
Website: An independent community project meeting the needs of local people through a range of social, environmental, educational, recreational and economic activities. There is a shop selling organic produce grown or reared on the farm. Website gives details of faculties and activities.
Contact: John Purkiss
Email: info@windmillhillcityfarm.co.uk
Address: Philip Street, Bedminster, Bristol BS3 4EA
Tel: 0117 963 3233
Fax: 0117 963 3252

Wing of St Mawes
www.cornish-seafood.co.uk

Specialities: Fish; cookery courses
Website: Fresh fish and shell fish from Cornwall despatched overnight to individual customers. Smoked fish from their own kilns is also available. There is a service for caterers which requires on-line registration. Export information is also given in French and Spanish. Illustrated recipes can be viewed on the site. Information is given on seafood cookery courses.
Links: Yes
Export: Yes
Email: enquiries@cornish-fish.co.uk
Address: 4 Warren Road, Indian Queens, St Columb, Cornwall TR9 6TL
Tel: 01726 861666
Fax: 01726 861668

Wino Organic Wines
www.wino-eu.com

Specialities: Wines; juices; cider
Website: The website allows internet visits to the three English vineyards and the one fruit winery that have been given the Soil Association organic certificate. It is also possible to buy on-line. There is a page on the benefits of organic produce.
On-line ordering: Yes
Credit cards: Most major credit cards
Contact: Allan Markey
Email: info@wino-eu.com
Address: 4 Station Road, West Hallam, Derbyshire DE7 6GW
Tel: 0115 932 0853

Wistbray Ltd
www.rooiboschtea.com

Specialities: Tea; rooibos tea
Website: Importer of Organic Eleven O'clock Rooibos Tea which is caffeine-free, low in tannin, high in anti-oxidants, and 100% organic. Website gives details and background of Rooibosch tea as well as names of stockists. Organic Dragonfly teas are also sold. There is a link for on-line ordering from www.goodnessdirect.co.uk.
Links: Yes
Contact: Bruce Ginsberg
Email: info@wistbray.com
Address: PO Box 125, Newbury, Berkshire RG20 9LY
Tel: 01635 278648
Fax: 01635 278672

Women's Advisory Service
www.wnas.org.uk

Specialities: Organisation; bodycare; herbal remedies; aromatherapy products; homoeopathy
Website: Offers scientifically based tailor made programmes for women through clinics in London, Lewes and Hove, as well as telephone and internet service. Lectures are given to the public and medical profession. The service also publishes self-help books. Website gives details of the many services available as well as an on-line shop for appropriate herbal and other products. There is a detailed health file and many information pages.
On-line ordering: Yes
Credit cards: Most major credit cards
Contact: Cheryl Griffiths
Email: wnas@wnas.org.uk
Address: The Old Coach House, Castle Ditch Lane, Lewes, East Sussex BN7 2QN
Tel: 01273 487366
Fax: 01273 487576

Women's Environmental Network
www.wen.org.uk

Specialities: Organisation; environment; nappies
Website: A national UK membership charity which provides information and campaigns on environmental and health issues from a women's perspective. Founded and run by women, membership is open to men and women. Campaigns include the encouragement of local organic food growing projects, waste minimisation (beyond recycling), health and the environment, sanitary protection and nappies. Co-organiser of Real Nappy Week every April and Healthy Flooring Network. Website details various campaigns, describes WEN's aims and successes and contains information on local groups, jobs, volunteering and support.
Links: Yes
Catalogue: Briefings & other publications available online or send stamped sae for details.
Email: info@wen.org.uk
Address: PO Box 30626, London E1 1TZ
Tel: 020 7481 9004
Fax: 020 7481 9144

Women's Institute Country Markets Ltd
www.wimarkets.co.uk

Specialities: Organisation; markets
Website: WI country markets provide an outlet for home produce in large and small quantities. The site gives information on how to become a producer, where to find market stalls, learn more about the produce, how to locate specialist products and other benefits. There are 500 plus weekly markets in the UK and there is a search facility by county on the site to help locate them.
Email: info@wimarkets.co.uk
Address: 183a Oxford Road, Reading, Berkshire RG1 7XA
Tel: 0118 939 4646
Fax: 0118 939 4747

The Woodcraft Folk
www.woodcraft.org.uk

Specialities: Education; organisation
Website: An educational movement for children and young people designed to develop self confidence and activity in society in order to build a world based on equality, friendship, peace and co-operation. The website provides a full introduction to the organisation, its activities and its outdoor centres. The outdoor centre brochure can be viewed or downloaded.
Links: Yes
Email: info@woodcraft.org.uk
Address: 13 Ritherdon Road, London SW17 8QE
Tel: 020 8672 6031
Fax: 020 8767 2457

Wooden Wonders
www.woodenwonders.co.uk

Specialities: Woodcraft; gifts
Website: Wooden gifts are made from Forestry Stewardship Council wood. Website gives details of the wooden pebbles and the woods from which they are made. There is an enquiry form for emailing for further details.
Contact: Noel Hardy
Email: info@woodenwonders.co.uk
Address: Farley Farm House, Chiddingley, East Sussex BN8 6HW
Tel: 01825 872691
Fax: 01825 872733

The Woodland Trust
www.woodland-trust.org.uk

Specialities: Organisation; woodland; conservation
Website: The UK's leading conservation charity dedicated to the protection of the native woodland heritage. Website has a search facility for locating woods in different regions. The Trust sets out its policies and has section headings on conservation with news and comment, briefings, action plans, agenda, parliamentary newsletter, ePolitix, consultation responses, publications and woodland threats. There is also general information on the trust and how to support and join it.
Email: enquiries@woodland-trust.org.uk
Address: Autumn Park, Dysart Road, Grantham, Lincolnshire NG31 6LL
Tel: 01476 581111
Fax: 01476 590808

Woodland Skills Centre
www.greenwoodworking.co.uk

Specialities: Activity holidays; holidays
Website: Activity holidays including chair making, wood-carving, coracle making, basket making and boomerang making. Website gives details of courses and shows what is involved in selected ones. There are details of accommodation available in the area, including camping.
Links: Yes
Catalogue: Brochure available
Contact: Tim Wade
Email: info@greenwoodworking.co.uk
Address: Greenwood Courses, The Church Hall, Llanafan, Builth, Powys LD2 3PN, Wales
Tel: 01597 860 469
Fax: 01597 860 469

Woodlands Farm
www.woodlandsfarm.co.uk

Specialities: Meat; turkeys; vegetables; box scheme
Website: A mixed farm with a herd of Lincoln Red cattle, organic bronze turkeys and sheep. It also produces organic vegetables and salads. The farm supplies both wholesale and direct to individuals. Website gives details of organic turkeys, box scheme, wholesale and recipes. There is a newsletter.
Links: Yes
Email: info@woodlandsfarm.co.uk
Address: Kirton House, Kirton, Boston, Lincolnshire PE20 1JD
Tel: 01205 722491
Fax: 01205 722905

Words of Discovery
www.wordsofdiscovery.com

Specialities: Children's books; board games
Website: Books to nurture children's personal growth. The mission is to bring books from around the world to promote children's holistic, spiritual and personal development. Topics include self-belief, tolerance, and spiritual and environmental awareness. There are also parenting books.
Links: Yes
On-line ordering: Yes
Email: sales@wordsofdiscovery.com
Address: Freepost LON7858, Leicester, Leicestershire LE5 6ZY
Tel: 0845 458 1199
Fax: 0116 262 2244

Worldwide Education Service
www.weshome.demon.co.uk

Speciality: Home education
Website: WES supplies tutorial based learning courses for children aged 4 - 13 years based on the National Curriculum of England and Wales. It also provides standard assessment testing service. The website gives a full overview of the services. There is an email form for requesting an enrolment form, fee sheet and further details.
Email: office@weshome.demon.co.uk
Address: Unit D2, Telford Road, Bicester, Oxfordshire OX26 4LD
Tel: 01869 248682
Fax: 01869 248064

G.R. Wright & Sons Ltd
www.wrightsflour.co.uk

Specialities: Flour; bread mixes
Website: Bread mixes including organic stoneground wholemeal. Website has information on the company and its products. There is a homebaking section which explains how flour is milled, provides recipes and tips and explains how bread mixes can successfully be used in bread machines.
Links: Yes
Email: homebaking@wrightsflour.co.uk
Address: Ponders End Mills, Enfield, Middlesex EN3 4TG
Tel: 0800 0640100

WWF The Global Environment Network
www.wwf-uk.org

Specialities: Organisation; conservation; resource
Website: The largest independent conservation organisation. The site has a search facility by words describing a concept or key words which will search for articles and facts on different topics on the site. There are also pages of facts and information, education and community, campaigns and support action.
Email: wwf-uk-supportercare@wwf.org.uk
Address: Panda House, Weyside Park, Godalming, Surrey GU7 1XR
Tel: 01483 426444
Fax: 01483 426409

Wyke Farms Ltd
www.wykefarms.com

Specialities: Dairy; cheese
Website: Dairy farm producers of organic cheese including organic cheddar. Website is available in English and French. It provides a history of Wyke, cheese information and tradition, the aims of Wyke Farms and details of the farm. There is a newspage and a map.
Contact: Richard Clothier
Email: sales@wykefarms.com
Address: White House Farm, Bruton, Somerset BA10 OPU
Tel: 01749 812424
Fax: 01749 813614

Xynergy Health Products
www.xynergy.co.uk

Specialities: Food supplements; wheatgrass juice; skincare
Website: Aims to provide the finest quality green, wholefood supplements from around the world as well as the best skin products. Index covers superfood supplements, Chinese herbal tonics, honey's propolis and pollen, aloe vera skincare, royal jelly skincare, haircare, bodycare and dental care. There is a gift service and on-line ordering.
On-line ordering: Yes
Catalogue: Leaflets
Contact: Sam St Clair-Ford
Email: naturally@xynergy.co.uk
Address: Elsted, Midhurst, West Sussex GU29 0PT
Tel: 01730 813 642
Fax: 01730 815 109

Yalding Organic Gardens
www.hdra.org.uk/yalding

Specialities: Seeds; gardening
Website: The website gives an illustrated description of Yalding Organic Gardens which are open to the public. Opening times and events are shown.
Address: Benover Road, Yalding, Nr Maidstone, Kent ME18 6EX
Tel: 01622 814650
Fax: 01622 814650

Yewfield Vegetarian Guesthouse
www.yewfield.co.uk

Specialities: Guesthouse; holidays; vegetarian
Website: Website provides details of accommodation, self-catering, the grounds, location, restaurants and tariff. There is an email booking request form. Concerts are held during the summer and information is given on the website.
Email: Derek.yewfield@btinternet.com
Address: Hawkshead Hill, Hawkshead, Ambleside, Cumbria LA22 0PR
Tel: 015394 36765
Fax: 015394 36096

Yoga Matters
www.yogamatters.co.uk

Speciality: Yoga equipment
Website: Company aims to supply good quality yoga equipment and the best value for money. All equipment comes with a no-quibble guarantee. Orders can be placed on-line, by phone or by post. Product range includes mats, blocks, belts, bolsters and publications.
Links: Yes
On-line ordering: Yes
Credit cards: Most major credit cards
Email: enquiries@yogamatters.co.uk
Address: 42 Priory Road, London N8 7EX
Tel: 020 8348 1203
Fax: 020 8341 9610

Yoga Therapy Centre
www.yogatherapy.org

Speciality: Yoga
Website: Website explains the nature of yoga therapy and how to try it. It also provides information on training and research. Books, yoga equipment videos and audio tapes are available for purchase.
Links: Yes
Email: enquiries.yogatherapy@virgin.net
Address:
Royal London Homoeopathic Hospital,
60 Great Ormond Street,
London WC1N 3HR
Tel: 020 7419 7195
Fax: 020 7419 7196

Yorkshire Game Ltd
www.yorkshiregame.co.uk

Specialities: Game; venison
Website: Wild venison, game birds, rabbits, hare, and pigeons from Yorkshire and the Scottish borders supplied to individuals and the catering trade. The website describes the product range which is available in larder boxes or dinner party packs.
Credit cards: All major credit cards
Contact: Sandra Baxter
Email: enquiry@yorkshiregame.co.uk
Address: 18 Leaside, Newton Aycliffe, County Durham DL5 6DE
Tel: 01325 516320
Fax: 01325 320634

Young People's Trust for the Environment & Nature
www.yptenc.org.uk

Specialities: Environment; sustainable development; organisation
Website: A charity which aims to encourage young people's understanding of the environment and the need for sustainability. There are many fact sheets written especially for young people and many of them are available from the website. The website also gives information on residential courses, residential holidays, school talks and school membership.
Links: Yes
Email: info@yptenc.org.uk
Address: 8 Leapale Road, Guildford, Surrey GU1 4JX
Tel: 01483 539 600
Fax: 01483 301 992

Youth Hostels Association
www.yha.org.uk

Specialities: Youth hostel; holidays; organisation
Website: Over 70 years of experience of providing value-for-money accommodation for backpackers, families and groups. The website provides a complete listing of hostels with a search facility and detailed information on individual hostels. There is a list of hostels with on-line booking facilities.
Address: Trevelyan, 8 St Stephens Hill, St Albans, Hertfordshire AL1 2DY
Tel: 01727 855 215
Fax: 01727 844 216

Zen School of Shiatsu
www.learn-shiatsu.co.uk

Specialities: Shiatsu; courses; practitioner training
Website: Website provides a general introductory overview of shiatsu. Courses available range from absolute beginner to full professional qualification. To help choose the right course there is a course management tutorial on-line. Details of certificates and diploma courses are given. There is also a 5 day intensive course. There is a list of recommended reading which can be ordered through a link.
Links: Yes
Email: info@learn-shiatsu.co.uk
Address: 188 Old Street,
London EC1V 9BP
Tel: 0700 078 1195

Index
by County

INDEX BY COUNTY

SCOTLAND

WALES

IRELAND

Domain Name Index

A

D

E

Domain Name Index

F

G

Domain Name Index

H

L

M

P

S

T

U

V

W

X

Y

Z